SECURITY

Theft Prevention, Security Development,
Fire Protection, Emergency and Disaster Planning,
and Guard Organization

WALTER M. STROBL
Vice President, Guardsmark, Inc.

INDUSTRIAL PRESS INC., 200 Madison Avenue, New York, N.Y. 10016

For my Wife, Carolyn

Library of Congress Cataloging in Publication Data

Strobl, Walter M   1921–
　　Security; theft prevention, security development, fire
　protection, emergency and disaster planning, and
　guard organization.
　　1. Industry–Security measures.　I. Title.
　HV8290.S85　　　　　658.4'7　　　73-16105
　ISBN 0-8311-1101-1

SECOND PRINTING

SECURITY

Contents

Foreword

Security is one of the fastest growing industries in the United States today. This includes not only the services offered to business by professional security companies, but also the individual protective plans in effect by individual firms.

The rapid growth of the industry began in the late 1950s and continued unabated through the '60s and into the '70s. This growth can be attributed to a number of causes, among them the tremendous proliferation of installations requiring security measures; a drastic increase in pilferage and stealing by employees; social unrest which has resulted in dissident groups turning to civil disobedience and violence to express their views; the inability of police and fire departments to keep pace with the demands being made upon them; and a growing concern for the safety of both employees and plant equipment.

It is estimated that dishonest employees are stealing at an annual rate of between $5 billion and $15 billion. No one knows exactly what the total is because of the obvious difficulty of compiling accurate statistics. But, it is estimated that pilferage and stealing are increasing at an annual rate of between 15 and 20 percent.

Growth of the industry has served to illustrate that putting guards at the gate and, perhaps, at other strategic locations on the grounds, is no longer sufficient to assure protection of property and employees. The growth has helped to point up the need for sophisticated application of security measures to fully utilize the vast array of hardware which has become available (including such items as closed circuit television and digital locks) and the need to help plant management to find and remove hazards which can endanger both employees and plant equipment.

Coincidentally, it has become apparent that the security industry and business must join hands to solve security problems.

Anyone who has ever had anything to do with planning security knows how many hundreds of details there are to consider if every possible breach is to be anticipated and closed. The vast majority of these possible breaches has to do with the frailty of the human being — his carelessness, his forgetfulness, or his intentional wrongdoing. A good security plan takes care of all these possibilities.

Beyond these, there are the fire hazards, material protection, emergencies arising from equipment failure, floods and other acts of nature, and the list goes on. But it is the human side — pilferage, organized theft, espionage, infiltration, sabotage, civil disturbance, the bomb threat, and even employee morale — which many times has presented the most perplexing security hazard.

Walter Strobl is very well qualified to write about all these subjects. His wide experience and background place him among the foremost authorities on security in the industry today. He has developed countless security plans for companies in this country and abroad.

The professional security man can use *Security* as his reference guide to determine the completeness of his own plan.

Even a beginner or amateur, with the help of this book, can plan a security system from the perimeter right into the very heart of the plant — not only to stop thefts, but also to guard against fire and to reduce hazards that threaten the safety and well-being of workers, equipment and material.

Ira A. Lipman, President
Guardsmark, Inc.

Preface

This book has been written for readers who have felt the need for a reference work devoted exclusively to providing practical answers to the many everyday and long-term problems encountered in developing, applying, and maintaining a physical security plan at the individual facility, whether it is an industrial plant, a trucking terminal, a warehouse complex, a food or garment distribution center, or any similar place of manufacture or business.

All the material is based on proven economic methods and procedures. Throughout the book I have avoided generalities and theories and have tried consistently to provide information which is very specific and which the reader should be able to put right into practice.

The emphasis is on (a) identifying and describing all the most commonly encountered industrial and business security hazards, (b) presenting the basic principles of modern and effective physical security, and (c) providing the proven practical solutions needed to eliminate or reduce the hazards.

If a physical security plan is to be effective and efficient, it is most important that all parts of the plan properly interlock, that loopholes are eliminated, and that the plan is fully "integrated." Experience has shown that these criteria are more likely to be met if the original study of the security requirements or needed improvements at a facility follow a well-tried sequence of steps, beginning at the facility site perimeter and progressing inward to cover the various areas of security in a certain logical pattern. The normal order of chapters in this book reflects this logical sequence of steps. If the reader works through the book by normal chapter order, he will be following automatically the steps which establish a correct sequence for conducting a physical security survey.

Besides theft and pilferage, losses to industry and business through fire, natural disasters, sabotage, and even bomb attacks are also of great significance. The sections on these subjects are likewise very practical and based on proven applications. Chapter 16, Emergency Evacuation and Disaster Planning, takes the form of an actual manual which could be readily applied at an individual facility. I believe this is a unique presentation which should facilitate early application at many companies (for permission to reproduce the manual material see page 195).

Likewise, I believe the extensive physical security checklist given in the appendix is also a unique presentation, and should prove a very useful aid to the reader in assessing and determining security conditions at a facility. However, the reader should fully digest the rest of the book before applying the checklist. Without this background, it is quite impractical if not impossible, to properly analyze present conditions and determine what changes are needed that will be effective, economical, and most important, workable.

The guard force under the direction of the security director is ultimately responsible for implementing the security program. The organization and deployment of the force is of utmost importance. It would be pure folly to organize the guard force and then attempt to develop the security plan. A complete survey and an analysis of the findings must be completed before a determination can be made as to exactly how many guards will be needed, how the guards will be deployed, what training will be required, and what their individual responsibilities will be.

Thus, this subject is covered in the last chapter of the book, as it is the final step in the natural sequence of events in the development of the security plan.

Over the years, many good friends advised and encouraged me to put my long experience in plant security into book form, and I would like here to thank them for their persistence. Without the patience, inspiration, and dedicated assistance of my wife, Carolyn, however, this book surely would never have been completed.

Special mention must also be made to those of my associates and friends whose invaluable professional advice and technical assistance was especially helpful and greatly appreciated; to Mr. Ira A. Lipman of Memphis, President of Guardsmark, Inc., who kindly allowed me to include certain material I had previously developed for the company as well as permitting me to use the company's technical library, and to Mr. Bedford "Bud" F. Joyner, Jr., of Integrated Communication Associates, also of Memphis, for his professional assistance in creating the drawings.

And, finally, my most sincere thanks to the various companies who have provided the many photographs and permitted their reproduction in the book.

CHAPTER 1

Consideration of Existing Conditions

An intelligent evaluation of the security requirements at any facility must start with a consideration of certain existing conditions. There are a number of preliminary steps to be taken and factors to be considered which, when defined, are then kept constantly in mind during the development of the whole security plan. These steps and factors include the following:[1]

(1) Before proceeding with a physical examination of the facility, it is of major importance to estimate, as accurately as possible, past losses sustained through theft and pilferage. This estimate should not only take into account actual known dollar values of stolen items but should include also values for the effect losses had or may have had in obtaining or sustaining maximum production. When the existing security conditions are being analyzed finally, and prior to making the final decision on the security plan to be effected, the cost of the program for security improvement should be balanced against the dollar savings the reduction in pilferage would accomplish and the value of increased availability of products to the customer.

(2) Before the physical plant study begins, the security director should obtain data on weather and particular natural phenomena peculiar to the locality where the facility is situated. Such phenomena as severe electrical storms, flooding, earthquakes, extreme

---

[1] A detailed Physical Security Survey Checklist which complements this and other chapters, begins on page 254.

wind conditions, or extreme heat or cold will have a bearing on the formulation of the final protective program, and special account may have to be taken when designing building structures or planning disaster and emergency evacuation procedures.

(3) A detailed study of the facility layout with reference to the site or plot plans should be made. This study will help to provide an overall picture of existing security arrangements and assist with analyses needed to develop and/or improve the security plan. It is vitally important to know the relative position of the buildings falling within the protected area and the geographical distances within the area to be protected; these distances and the time and distance factors are vital if security patrols and/or security hardware are to be successful in the overall security program.

(4) To insure that there is no interference with the objectives of the facility, and to avoid impeding operational efficiency, the security director should familiarize himself with all the operational flow plans. This will be important in deciding what security measures to adopt in areas where theft and pilferage are possible.

(5) The security director should be aware of the economic status and sociological conditions existing in the immediate area of the plant location. For example, if the facility is situated in or adjacent to an area where the crime rate is high or growing, it may be necessary to take account of this fact when developing the type and extent of protection installed at the perimeter.

(6) The past and present labor relations at a facility have an important bearing on security and this fact should be taken into account. Where management/labor relations are poor, an atmosphere may exist which is conducive to pilferage and theft. When mutual loyalties are at a high level, there should be less pilferage. The outlook of the community as a whole, and the view the community takes of a particular company may have a similar bearing on security risk, and must be taken into account when developing the type and extent of protection installed at the perimeter. Also, products of one company may be more desirable for theft than those of another. When deciding on security measures, the security director should keep this point in mind.

(7) Where union agreements exist, it may be desirable to discuss some aspects of the security program, or the plan as a whole, with representatives of the local labor organizations. This decision should be based on past and present relationships with labor officials and the attitudes of union employees. Whether or not the facility is unionized, the protection plan's effect on employee morale must be carefully examined. It is extremely important that this evaluation include a discussion with line management to give them the facts and gain their impressions.

(8) Employee statistics are another important consideration in establishing a security plan. The number of employees in each category—administrative employees, supervisory employees, hourly employees, etc.,—and the number of shifts, and the times shifts arrive and depart all have a direct bearing in establishing an employee identification system and control of persons within the protected area or entering or leaving it. Any contemplated increases or decreases in employee population and the percentage of turnover of employees in any given category should be taken into account.

(9) The legality of implementing certain aspects of the security plan must be researched to insure that local ordinances or state laws are not being violated.

(10) The location relative to the facility of law enforcement agencies and fire departments, the reliability of these departments, and their response and assistance in emergencies, as indicated by past performance, should also be considered. They could have a direct bearing not only on the measures and techniques employed in the overall security program but also on plans formulated for coping with emergencies and civil disturbances.

(11) The final consideration, yet one that is actually most vital to the success of the program, is the method by which the program will be implemented. The implementation of the program must develop the acceptance and cooperation of the employees as a group. The degree of success of a new or revitalized security program can be directly related to the employee educational program which is launched prior to the program's initiation. Therefore, it is

very important to establish effective communication between management and employees.

In order to gain the employees' wholehearted acceptance and support of the program, management will have to subject themselves to many of the restrictions imposed on the employees. This subject is so important that an entire chapter (Chapter 20) has been devoted to security education of employees.

CHAPTER **2**

Recognition of Security Hazards

A *security hazard* may be defined as any act, omission, or condition which could seriously breach the protective system and result in a compromise of company secrets, a loss of company property, or injury to personnel. Before a security hazard can be eliminated or reduced, however, one must be able to recognize it.

It is often said that it is easier and less expensive to prevent or deter illegal or unsafe acts than it is to detect and apprehend individuals after they have committed them. This is true not only as regards direct monetary considerations but also in respect to insuring continued maximum production. Most undesirable incidents which occur in an industrial facility have a direct adverse effect on employee morale and a subsequent slackening of productivity.

Sometimes it is necessary, in order to reduce or eliminate theft and pilferage hazards, to employ an undercover investigator to identify the offenders. But such situations usually arise where no security program, or an ineffective one, previously existed.

Therefore, an effective industrial security program should deter, reduce, or eliminate all existing security hazards. In some instances, it will be physically impossible to eliminate certain hazards to security; when such conditions exist, the effort should be directed toward reducing the calculated risk to an acceptable level.

Security hazards can be divided generally into two major categories: human hazards (undoubtedly the most troublesome to the security director) and natural hazards.

5

HUMAN HAZARDS

Through the use of security aids, techniques, and devices the human hazards can, usually, be effectively dealt with. These hazards can be defined as: theft, pilferage, carelessness, disloyalty, dissatisfaction, sabotage, and espionage.

The order of frequency in which these incidents occur is, in most instances, the same order as that listed above, with pilferage vying with theft for first place. Admittedly, the relationship between these two illegal acts is quite close. However, thefts normally involve greater quantities of property over shorter periods of time than pilferage by employees, which may be committed almost routinely.

Pilferage and outright theft are by far the most common and most annoying of the human hazards. In analyzing the actions of the thief and pilferer, it should be realized that, in these instances, two specifically different groups can be identified, and both can operate within the same facility at the same time.

The Casual Pilferer

The casual pilferer, who normally has no definite pattern or system to his actions, steals small items only when the opportunity arises. When apprehended, he will usually be found to have taken items of company property for his own use, for a friend, or for a quick resale. He usually does not steal on a regular basis and will not steal unless he is sure the possibility of being apprehended either does not exist or is minimal. The casual pilferer is usually caught through such security measures as unannounced lunch box, package, or purse inspections.

The Systematic Pilferer

The systematic pilferer removes large quantities over a short period of time. He develops a definite plan of action taking into account his firm's security measures. His plan includes other individuals within the plant and a means of disposing of the stolen property once he has successfully removed it. Obviously, the systematic pilferer is more dangerous than the casual one, since he operates on preplanned methods, selecting items of high value which are small enough to insure that the chances of detection when he removes them are minimal.

The systematic pilferer may work alone, but usually he works in groups of various sizes which may include fellow employees, visitors to the plant, applicants at the personnel office, truck drivers serving the plant, and other individuals actually authorized to be within the protected area for specified periods of time. His *modus operandi* involves individuals not employed at the facility who will receive the stolen property once it has been removed.

There are several methods of concealing items stolen from an industrial plant or business and literally hundreds of variations of these methods, but only the simplest and those most often used will be dicussed here.

Moreover, items may be removed without concealment. In some instances, items stolen from a particular type of manufacturing facility may be removed in plain view of the guard, without his knowledge. As an example, the manufacturer of sweaters, skirts, trousers, or shoes could be sustaining losses because these items can easily be removed from the property by wearing them in full sight of the guard. A guard should be familiar with the items manufactured in the plant and alert enough to discover them being stolen.

Substitution is another ploy. In one clothing firm, the retail price of sweaters ranges from $25.00 to $75.00. Management has set a policy that usually enables employees to purchase this merchandise at a discount, at an employee sales store, either on a regular basis or periodically during the year. An employee who has bought one of the less expensive sweaters of a particular type or design wears it to work the following day. Sometime during the day he discards this sweater and dons one of the more expensive styles in a similar color or style. His chances of being apprehended as he departs are practically nil. He will return to the plant on some later day, will recover his cheaper sweater and walk out of the plant past the guard, knowing that he can prove ownership. At a later date he will repeat his illegal performance.

Methods Used to Pilfer

Probably the most common method of removing a pilfered item is to conceal it in the employee's clothing, lunch pail, packages, or on vehicles. During the cold weather months, when the employees wear additional clothing, concealing items on the person becomes that much easier. It is difficult, if not impossible, to detect small items hidden in pockets and under outer clothing.

If employees are required to leave their outer clothing in lockers rather than carry them to their work stations, pilferage will be substantially eliminated or reduced. The lockers should be subjected to periodic, unannounced inspections by the security force and a member of management. To facilitate locker inspections, management should furnish the locking device and retain a key; if locks are placed on lockers by the individual employee, management must be furnished a key.

The practice of secreting stolen items in lunch pails, purses, and briefcases can usually be eliminated by periodic inspections at unscheduled times by the guard force. When the guard force examines these containers, a member of management should be available to initiate disciplinary action in the event the guard discovers unauthorized company property being removed.

If employees are authorized to carry packages into the facility without checking them, or if they are authorized to carry packages out of the plant from, for example, the company store, and there is no system for control, rest assured the opportunity will be taken to pilfer. If management authorizes packages to be removed from the plant, it must institute a package pass system.

Vehicles allowed within the premises are another hazard. Employees authorized to park their personal vehicles within the perimeter barrier will very probably use these vehicles, at one time or another, to remove items pilfered from the company. Trash and garbage removal operations are tailor-made for the removal of large items or a large quantity of small items from the plant. These refuse removal operations must be closely supervised and controlled by the guard force.

Where building walls form a part of the perimeter barrier, and these walls are breached by windows or doors, additional precautions must be taken to insure that pilferers do not throw items outside, either for later recovery or to a confederate waiting nearby.

A variation of the above method is employed by pilferers who remove items from a building and secrete them in the area between the building and the perimeter barrier. When the opportunity presents itself, they move the items closer and closer to the barrier, until, eventually, they throw them either over or under the barrier, to be recovered later by the employee, or immediately by a confederate waiting outside.

Transports and railroad cars moving in and out of the protected area must be closely scrutinized by the security force to effectively deter pilferage and theft accomplished by this means.

These are merely some of the most common methods employees use to steal and pilfer. The resourceful thief will devise his own methods.

One systematic pilferer stole thousands of dollars worth of small precision tools during the hours of darkness by tying them to inflated balloons and releasing them when the wind direction and velocity were most advantageous. His confederate, some distance outside the perimeter barrier, recovered the stolen items by shooting down the balloons with an air rifle.

For many years another employee delivered his family's supply of meat by placing the amount called for in that day's menu in a plastic bag and dropping the bag in a stream flowing through the protected area. This same stream flowed through the employee's backyard approximately one-half block from the plant, and the meat came floating by regularly at 4:00 P.M., in good time for his wife to prepare their dinner.

Remember that the methods used today to effectively deter theft and pilferage may be obsolete tomorrow because of changes in the methods employed and the ingenuity of the thief in devising new methods. Therefore, any security measures which are adopted must be updated to insure that the degree of security remains high.

Carelessness

The careless employee is as great a security risk as the pilferer. This individual will leave doors open when they should be locked, will allow easily pilfered items to remain unsecured, will drop cartons of the packaged finished material, thereby exposing the smaller boxes to theft and pilferage, and will also be a real safety risk to himself and others.

Since physical security is only one part of the overall plant loss-prevention program, the careless individual must be identified and removed from the area where he has become a high risk or his dismissal might have to be considered.

Disloyalty

Disloyalty in an employee usually occurs because he is disgruntled with his work or, in one way or another, has been disciplined by a member of management. In some instances, employees have been bribed by outside individuals; however, this is the exception rather than the rule.

The disloyal employee can most easily be identified by close observation of his actions and productivity. The degree of disloyalty can usually be measured in direct relation to the individual's personality, attitude, and character. As an example, the severity of the incident he perpetrates can usually be related to the severity of the discipline he has received. Experience has proved that the disloyal employee will more usually vent his anger and frustration by words rather than by actions. Therefore, he can quite easily be identified. Once he has been identified, immediate remedial action through counseling should be initiated.

Dissatisfaction

The dissatisfied employee develops an attitude similar to that of the disloyal employee; probably the greatest difference lies in the fact that the dissatisfaction results from conditions the employee thinks are unjust rather than from the actions of one or more individuals.

The dissatisfied employee will harbor his feelings for long periods of time until, eventually, his smoldering resentment can no longer be retained and it surfaces. Often, as a result, the employee takes some physical action against the employer. If an employee is dissatisfied because he didn't receive an expected pay increase, he might turn to regular pilfering to make up for what he considers the loss of pay due him.

Espionage

Fortunately, the industrial espionage agent is relatively rare. However, the threat should never be underestimated, for the espionage agent is a very real and very dangerous adversary with great skill in deception and cunning. This type of individual has usually had extensive training in espionage methods and, through his assumed personality, he is highly effective in extracting information of value which relates to his target. His work, today, is more than likely directed toward securing company proprietary information and secrets. The results of his work are not always immediately apparent but they usually have a far-reaching adverse financial affect upon the victim company.

The security force's effectiveness in combating the espionage agent lies in its control duties. It must insure that any individual not properly documented does not gain access to the protected area.

Sabotage

The saboteur is more likely to be encountered or confronted by the security force than is the espionage agent. The saboteur, most often, is an employee who has or had authorized access to the protected area.

The words "sabotage" and "saboteur" were coined as a direct result of industrial plant employees' actions against their employer and their country. During World War I, great numbers of French industrial workers were still wearing wooden shoes that, in French, are called *sabots*. Those employees who harbored grudges against their employer, and those who were not in agreement with national policy, would drop their wooden shoes "accidentally" into machinery, causing destruction and slowing down the manufacturing process.

Most sabotage committed against industry today can be traced directly to the actions of an employee, usually a disloyal or dissatisfied one. However, these acts of sabotage are usually confined to more specific periods of unrest.

Sabotage at an industrial complex can be expected to occur prior to or during a labor strike. This type of sabotage is designed to prevent manufacturing processes from continuing. Strike saboteurs can easily be identified by line supervision through close scrutiny of their actions.

Another form of sabotage that may be encountered by the modern industrial facility is the illegal actions of activists against the facility. This type of attack may be implemented merely because of the facility's real or assumed role in producing war material. The defense against this individual or group of individuals is comparatively simple, providing an effective overall physical security program is in effect. Psychological conditions leading up to acts of sabotage are usually evident for a sufficient period of time to permit management and the security director to diagnose them and to strengthen the plant defenses.

NATURAL HAZARDS

The second major category of security hazards is referred to as natural hazards because they are acts of nature which may occur daily or are peculiar to a given locality. They may be divided into seven categories: heat and cold, darkness, fires, explosions, floods, tornadoes and hurricanes, and earthquakes.

Natural hazards are considerably more difficult to deal with than are human hazards. They are an immense threat to the security of an industrial facility; but, obviously, there is nothing that the industrial security force can do to prevent or deter these acts of God from occurring.

The security force's function when these phenomena occur, and the effectiveness of their action, must be directly attributed to their training in performing their functions. Security forces and their training must be considered an integral part of the security plan. In many instances, the only individuals present in the facility when these disasters occur are some members of the security force. Reliance upon them, individually and collectively, to implement immediate emergency procedures is mandatory. Retraining and review of their part in implementing emergency procedures must receive the highest priority in the security force training program.

Heat and Cold

Heat is a natural hazard to physical security because when the weather becomes exceedingly warm, windows and doors may be left open merely for ventilation purposes. Obviously, open and unprotected windows and doors afford the opportunity for an intruder to enter the plant or building to commit his illegal acts. Open windows and doors in the building are also a great temptation and an even greater opportunity, for the pilferer to remove the items he has stolen from the plant.

Therefore, during extremely warm weather when the windows are left open they must be further guarded. The security force must be extremely alert to insure these openings are not used to perpetrate thefts.

Cold is classified as a natural hazard because during cold weather employees wear heavy outer clothing when going to and from work. This gives them the opportunity to hide items in the pockets and conceal items under heavy outer clothing. As previously discussed it is difficult if not impossible to determine whether or not the employees wearing these heavy outer garments have hidden small items on their person. In most instances it is illegal to search the individual. If a search of an individual is conducted it should be done by a member of plant management and not left to the discretion of a member of the security force.

Additionally the advent of cold weather extends the potential threat to the security of a facility because of the use of heating systems and boilers. As an example, a heating system malfunction may cause freezing of the automatic sprinkler system rendering it useless. During the period of the year when heating systems are started up the potential of fires and explosions is increased. Freezing weather also increases safety hazards including the personal safety of individual members of the security force, because of icy walkways and slippery stairs that are traveled during guard patrol.

Darkness

Nightfall brings additional hazards to plant security. The individual intent upon stealing or committing acts of sabotage or vandalism will obviously take advantage of the hours of darkness to perpetrate his crime if it suits his purpose.

The only method of combating or reducing this natural hazard is through the proper installation of a protective lighting system. Such a system inside and outside the facility will deter the intruder or thief. Thieves do not like to work where sufficient light exists to provide possible detection of their illegal actions.

The protective lighting system has a most important function in the overall security plan of the facility and without it the perimeter barrier would be most ineffective. A perimeter barrier without lighting would expose the entire facility to surreptitious attack.

Chapter 5 deals with the protective lighting system and its role in the overall security plan.

Fires and Explosions

Fires and explosions are classified as natural hazards rather than human ones even though the conditions that cause such incidents may be closely related. The greatest percentage of fires and explosions experienced by industry are the direct result of violent storms, floods, earthquakes, and intense heat. In some instances improperly compounded chemicals may result in spontaneous natural conflagration.

Prevention and protection initiated to effectively combat fires and explosions cover volumes of text and information. The National Fire Protection Association, 470 Atlantic Avenue, Boston,

Massachusetts, is the accepted authority and issues an annually up-dated publication referred to as the NFPA Standards. These ten volumes should be in every industrial security director's reference library.

Chapter 13 is devoted to fire protection and prevention as it applies to the responsibility of the plant security director and the plant security force.

Floods, Tornadoes, Hurricanes and Earthquakes

These natural phenomena obviously cannot be controlled. There-fore if any of them is apt to occur in the area where the facility is located, due account must be taken in the design of buildings and structures, and in developing emergency plans.

As an example, if a facility is built in an area where flooding may be probable, even though it occurs very infrequently, consideration should be given to erecting dikes in those areas where flooding is likely to occur.

If the facility is located in an area where tornadoes, hurricanes, or earthquakes are prevalent the architect must take these destruc-tive forces into consideration when planning the structures. For example, in areas where tornadoes or hurricanes are likely to occur, glass exposures should be reduced or a means of covering peripheral windows during imminent danger should be planned. In areas where tornadoes are likely to occur the emergency plan most certainly will have to provide areas within the facility where em-ployees will assemble when the tornado "warning" is announced.

Criticality Assessment

The ability to recognize security hazards must include an assess-ment of how critically placed the facility as a whole is, and how vulnerable certain facilities inside the protected area are. Critical areas or installations at an industrial facility may be power distri-bution locations, fuel supplies, specific items of machinery, or the data computer center.

In order to establish the criticality of a facility, or an area within the facility, a careful evaluation must be made. Generally speaking, the criticality of an area or a specific peice of machinery can be measured by its overall contribution to sustaining successful operation and maximum production. The maximum effort should be directed to protecting those areas which are critical to the continued successful operations of the plant as a whole. As an example, delicate instruments should obviously be given greater protection than an empty railroad tank car.

Items of machinery, buildings, or equipment that have a high capability of self-destruction, or the capability of causing serious damage to other machinery or equipment, certainly require a higher priority in the assessment of the protection plan.

The basic unit involved in planning to minimize damage and to insure rapid resumption of operations is known as the "functional area." The functional areas which warrant primary protection can be analyzed by identifying these two factors: (1) relative importance to overall production or continued operation; and (2) relative vulnerability to damage of the building, the area within the building, the machine, or other equipment.

When considering the functional areas that have been given priority because of their importance to overall production, those appearing nearest the top of the list should receive first priority. The criticality of these areas should then be determined by a study comprising the following categories:

Group 1. All of those installations whose loss would cause an immediate stoppage of production because production equipment or parts would be lost

Group 2. Those areas or equipment where loss would reduce production or operations because of a partial loss of productive equipment or parts

Group 3. Those areas where any loss would not have an immediate effect on production or operations but whose loss would require additional manpower to maintain their functions

Group 4. Those areas in which a loss would not have any direct effect on production or operations.

Vulnerability Assessment

After establishing the criticality in those functional areas in the groups above, the groups will then have to be ranked according to their *vulnerability* to damage. As production or operations change, the grouping will have to be revised to maintain the proper ranking within each group.

This procedure establishes the criticality and vulnerability of certain functional areas within the industrial complex. It also establishes an order of priority for these functional areas based upon the effect on production and continued operations, should damage or destruction occur.

When this assessment has been completed it will be simpler to make decisions as to the degree of security that must be established to adequately protect these areas, functions, or installations. Obviously, the degree of security required to adequately protect those functions at the top of the list must, of necessity, be greater than the security furnished for those functions at the bottom of the list.

To cite another example of establishing the criticality and vulnerability of a particular area, assume that an industrial plant manufactures 85 percent of the electricity required to sustain operations. The remaining 15 percent furnished by a public utility company is the maximum amount of power that can be expected from this source.

The power plant is located some distance from the main facility because of the availability of water, and this location is outside the perimeter barrier protecting the facility.

The criticality study places the power plant at the top of the priority list because, regardless of any other critical installations, equipment, or machinery, if the power plant were damaged or destroyed, productivity would be immediately reduced or stopped altogether. Therefore, the relative importance of the power plant has been established. The loss of this installation would cause an immediate stoppage of production and it would fall under Group 1 mentioned earlier.

Since the power plant is located outside the protected area at a remote location, its relative vulnerability has also been established. The power plant will therefore be afforded top security.

CHAPTER 3

Perimeter Protection

With the knowledge accumulated as a result of research into those areas discussed in the preceding chapters, the physical examination of the facility can now begin. Remember, however, that as the physical examination of the facility continues, the considerations already studied must be kept constantly in mind to insure that carrying out the final protection plan will secure the facility without interrupting its objectives.

In the overall physical security plan which will eventually be set up for the protection of the facility, four definite physical lines of defense or protection must be established. These lines of protection are established by physical means, with the avenues of ingress and egress controlled by a trained security force. The four lines of protection are:

1. *Perimeter Barrier.* This is the barrier normally installed on the established property line of the facility; or, if the real estate owned is substantial in size and only a portion of it is occupied by the facility, the perimeter barrier is established at a sufficient distance from the buildings to insure that adequate outside areas for operational purposes are available.

   The facility site or plot plans must be studied to establish the perimeter line.

2. *Area Security.* The area considered here is that real estate between the perimeter barrier and the buildings comprising the facility. Security of this area is established through security force patrols, installation of protective lighting systems,

organization of outside storage areas, and properly engineered employee parking lots. Physical security devices and closed circuit television surveillance systems are also often used in establishing the degree of security required.

3. *Peripheral Walls of the Building.* The walls of the building are considered the third line of protection, whether or not they form a part of the perimeter barrier or lie within the protected area of the barrier. Certain security measures must be adopted to protect the openings created by doors and windows which breach the peripheral walls of the building. This must be done whether they open into the protected area or outside the protected area. Windows opening outside the protected area may need a higher degree of security than those opening inside the area.

4. *Areas Inside the Building.* The fourth line of protection consists of tailoring security measures to protect certain processes, materials, or activities within the building. These areas will already have been defined through the criticality and vulnerability study and analysis.

This chapter will examine and discuss the first line of protection: the perimeter barrier.

Functions of the Perimeter Barrier

If the barrier is to be effective and actually provide the first line of protection, it should furnish at least the following functions in implementing the overall security program:

1. It must present a physical and psychological deterrent to innocent entry.
2. It must be substantial enough to delay intruders and attackers and assist in their detection and apprehension if the barrier is breached.
3. It must provide for an effective and economical deployment of security forces.
4. It must provide an effective means of directing the flow of personnel and vehicles through the established control points without creating undue delay, or disrupting operations.

Construction of the Barrier

The standard perimeter barrier normally consists of an industrial-type chain link fence meeting certain standards and specifications and often in accordance with those developed by the United States Defense Department early in World War II to increase the physical protection of the facilities then referred to as "defense plants." These standards and specifications are as follows:

1. The size or gauge of the wire used in the production of the chain link fence should be number 11 or heavier.
2. The chain link portion of the fence should be at least 7 feet in height.
3. It should contain an overhang or "cleavage" consisting of 3 strands of tightly stretched barbed wire angled outward and upward, away from the property being protected, at a 45-degree angle from the horizontal.
4. The mesh openings should not be larger than 2 inches square.
5. With the barbed wire cleavage installed at the top of the chain link fence, the total height should be 8 feet.
6. The fence should be extended to within 2 inches of firm ground and should be installed below the surface if the soil is sandy or easily shifted by weather.
7. The mesh fence must be drawn taut and securely fastened to rigid metal posts set in concrete. Additional bracing is necessary and should be placed at fence corners and gate openings.
8. Culverts, troughs, or other openings should be provided where necessary to prevent washouts under the barrier. If these openings are larger than 96 square inches in area, they must be provided with additional protection.
9. Openings in the barrier should be kept to a minimum consistent with operational requirements, because each gate established on the barrier creates a weak point. Even when a substantial locking device is used to secure the gate, an attack upon the barrier can be directed toward one point—the lock. Otherwise, numerous points at a given position on the barrier must be successfully attacked to create an opening.

10. Where buildings, trees, and utility poles are within 8 feet of the barrier, the barrier should be increased in height or the protection otherwise increased.

In some instances, masonry walls may be constructed to form all or a portion of the perimeter barrier. As an example, in some instances it may be objectionable to install an industrial-type chain link fence across the front of the property which houses the facility's offices. Fig. 3-1 shows a perimeter barrier consisting of a chain link fence and masonry wall.

If a masonry wall is used as perimeter protection, it should be eight feet in height and constructed of a strength equivalent to that of the remainder of the barrier. Because a masonry wall is somewhat easier to scale than a chain link fence, and since an outside attacker cannot be observed from within, means must be devised to make it extremely hazardous to scale the wall. This can usually be accomplished by imbedding pieces of broken glass or wrought-iron pointed barbs on the top of the wall, or by merely installing barbed wire cleavage angled outward and upward, similar to that used on the remainder of the barrier.

The installation of masonry walls to increase the protection of critical and vulnerable areas, materials, or installations, should also be considered. For example, it may be desirable to increase the physical protection of a railroad tank-car while it is being loaded or standing within the protected area, if it contains noxious chemicals which are injurious to personnel, or might even cause fatalities if the tank car ruptured. This means of protecting tank cars has been installed by chemical companies which were considered prime targets of dissidents who might possibly rupture the tank cars by firing at them with high-powered rifles from some distance outside the protected area.

Masonry walls have also been constructed to protect transformer banks. They have proved highly successful in the protection of these installations.

A survey was conducted of a large electrical producer and supplier in Venezuela, in 1963, when overt actions of foreign and national dissidents were at their height. Among the problems encountered was the protection of some fifty transformer banks, a great number

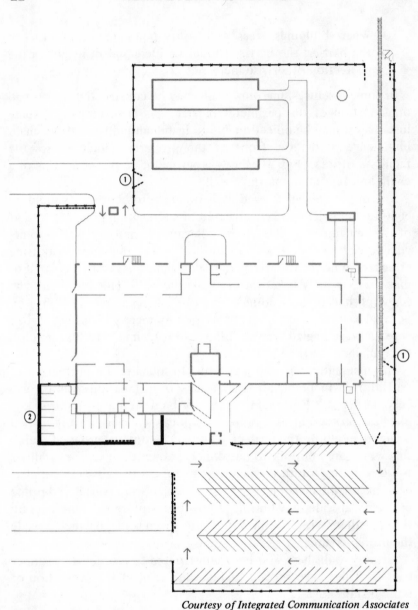

Courtesy of Integrated Communication Associates

Fig. 3-1. A perimeter barrier consisting of an industrial chain link fence and sectioned masonry wall (2), completely surrounds the facility. Note that the employee parking lot is outside the protected area. Gates (1) are installed where needed for operational and emergency purposes only.

of which were located in the rural mountainous and, for the most part, unpopulated areas. The attacks upon these transformer banks consisted primarily of gunfire from speeding automobiles. The flat projectory fire directed at the transformer banks would rupture the oil-cooling systems and, obviously, cause damage to and destroy electrical components.

Because of the extreme heat and the necessity of maintaining a continuous flow of air to assist in cooling the transformers, it was impossible to construct a solid masonry wall of sufficient height around the transformer banks. However, physical protection of some type had to be established, for it was uneconomical, unsafe, and unwise to attempt to protect each of these locations with armed manpower. It was decided to protect the transformers with masonry walls constructed in sections. The sections were approximately two feet apart and overlapped one another with approximately an eighteen-inch opening at the bottom of the wall. This type of construction allowed a continuous circulation of sufficient air, yet created a solid, continuous masonry wall for protection against flat trajectory fire.

Building Walls

When building walls form a part of the perimeter barrier, the windows and doors opening onto the perimeter must be further protected if they are larger than 96 square inches in area or less than 18 feet from ground level and less than 14 feet from any structure outside the perimeter.

The first consideration must be given to the strength in the construction of the building walls. If the strength of the walls is equal to or greater than the strength of the remaining barrier, the security effort is then directed to the windows and doors.

Windows can be adequately secured by placing chain link screens, gratings, or metal bars over the window openings. If fire departments are likely to use windows as points of entry, the additional protection should be installed so that it is hinged on one side and securely fastened with a padlock to the other. This type of barrier can be easily and quickly removed to create an opening for the firefighters.

Doors breaching the perimeter walls can be protected in a number of ways, depending upon their particular use. If they are created only for emergency exit purposes and are of solid construction, they can be secured with a local alarm to deter unauthorized use from within; or they may be electrically supervised by installing a simple electromagnetic device which is monitored remotely at a security force post or other location. This type of door should have all hardware removed from its exterior, with the exception, perhaps, of a key-operated locking device.

It is wise to reexamine operations in the immediate area of door and window openings to determine whether or not they can be eliminated entirely. Often the windows and doors were initially created for a specific purpose and, because of changes in the operational flow, they are no longer needed. In such cases, the openings may be permanently closed by installing bricks or cement blocks. This application could be advantageous from two viewpoints: (1) the additional degree of internal security established; and (2) protection against damage or fire bombings in the event of a civil disturbance.

If a building wall does furnish a part of the perimeter protection and the chain link barrier is anchored to the wall of the building, the height of the fence immediately adjacent to the building must be increased at least six feet and extended outward from the building at least eight feet, creating a fan-shaped barrier immediately adjacent to the building.

In facilities of earlier construction, coal chutes often breach the peripheral walls of the building to allow trucks to deliver coal without entering the protected area. If the coal chutes are no longer used, they should be permanently secured, either by locking the metal door in place from inside the building or by closing the opening with brick or cement blocks.

Doors with glass windows which open onto the perimeter barrier and are located in areas somewhat remote from normal activities should have the same protection as that afforded windows in the same area.

Perimeter Gates

Openings in the perimeter barrier should be kept to the minimum consistent with operational needs. However, the size of the area being protected, the roadway approaches to the facility, and the location of the public fire department servicing the facility may require that additional openings be created.

Openings created on the barrier to facilitate the entry of emergency equipment and vehicles will normally consist of double-swing, manually operated, vehicular-type gates. These gates should be secured with a substantial locking device, usually a padlock, because of their infrequent use. The responsibility for manning the gates must be clearly defined in the emergency and disaster plan procedures.

Quite often, the key to the padlock securing an emergency gate is issued to the fire department to facilitate its entry in an emergency. This is unwise; key control is immediately lost, since this key could easily be compromised. It this procedure must be followed, the locks on this type of gate should not be included in any submaster locking system established at the facility. The hazard may be reduced by retaining the key and notifying the servicing fire department to be prepared to destroy the lock to gain entry.

If all efforts in this direction fail, a railroad-type seal or a wire and lead seal should be applied to the gate, so that at least after-the-fact knowledge of whether or not the gate is being used surreptitiously can be obtained. Perimeter barrier inspections are most important.

There are several types of gates which can be installed on the perimeter barrier. The type selected will depend on the size of and purpose of the openings.

Personnel gates should be approximately five feet wide to facilitate a single-line entry or exit. Personnel gates normally establish employee and personnel control points and, therefore, are manned by the security force while used operationally. These gates are manually operated.

Vehicular gates consist of single-swing or double-swing gates; rolling gates with large casters installed on the bottom; cantilever gates, which operate by movement on an overhead tract; or overhead gates, which operate like overhead doors. The type chosen will depend upon the activity established in each specific area. Any type of gate can be electrically operated and remotely controlled electronically from a security post on the property.

Where a vehicular opening on the perimeter barrier is of sufficient size to permit ingress and egress of vehicles at the same time, and this point is manned by only one security guard, an additional barrier to assist in control could be considered. This barrier would be used only during operational periods and only for control purposes. Normally, a drop-type bar barrier is used and, depending upon the amount of vehicular activity, can either be electrically or manually operated. This control is often effectively used when vehicular traffic is heavy and the controls established require each vehicle to be registered in and out of the protected area.

The installation of these restrictive devices will insure that vehicles infrequently servicing the facility do not move in or out without proper documentation while the security guard is involved in registering another vehicle.

Railroad gates should be attended or under observation of a security guard while switching is in progress. They should be locked in a manner similar to that used for the remainder of the gates on the perimeter.

Railroad gates are often secured with two pieces of chain, with a "railroad-type lock" and a "company-owned lock" securing the two pieces of chain in place. This is done to enable the railroad employees to open the gate with their lock when switching is to be accomplished.

The type of lock used by railroads throughout the United States is a *lever lock*, referred to by railroad people as a "switchman's lock." Not only can it be defeated merely by striking the body of the padlock a sharp rap, but these locks are keyed alike throughout the entire United States. They furnish no security whatsoever; even if they did, if the railroad retained possession of a key, the security value of the locking device would be negated. Therefore, every effort must be made to include control of the railroad gates in the functions of the security force.

Miscellaneous Breaches of the Barrier

Infrequently, and usually confined to those facilities of an earlier construction age, conditions will be found which breach the perimeter barrier. Although these areas may not be in operational use at present, they do create very real weak points in the overall protection of the perimeter barrier.

1. *Sidewalk elevators* should be secured from beneath the opening cover by an adequate locking device.
2. *Utility tunnels* should be secured by installing doors and adequate locking devices to secure them when not in use.
3. *Operational tunnels* used in the more modern complexes must be secured in a manner similar to that used to secure operational peripheral doors which breach the barrier.
4. *Storm sewers or storm doors*, when larger than 96 square inches, must be further secured by installing a locking device on the manhole cover to secure it in place or installing a barrier across the discharge opening outside the perimeter barrier. Barriers installed over the discharge end of a storm sewer must be removable and should be inspected for clearance after every rainstorm.

In the Venezuelan operation previously referred to, the utility openings covered by a manhole cover outside of the protected area created quite another problem. These openings housed all of the electrical cables leading into and out of a particular transformer bank. The transformers were located within the city of Caracas and were protected by a solid masonry wall approximately 12 feet in height on all sides. The masonry walls were constructed when the transformers were installed several years previously, at a distance from the transformers great enough to provide proper cooling. These were being attacked from automobiles by dissidents who merely drove onto the sidewalk and threw their fire bombs blindly into the protected area. More often than not, their attacks were successful.

To establish protection for these sites, lumber was used to create a "decking," a rooflike installation over the entire protected area. "Chicken fence" wire, commonly used in the United States on poultry farms, was imported and installed by nailing it firmly to the lumber structure. The next two or three fire bomb attempts resulted in the bottles and glass jugs of jellied gasoline rolling off the taut wire and breaking and igniting on the street.

The dissidents soon found an answer to this protection by merely opening the manhole covers outside the protected areas and throwing their fire bombs into the service opening. This operation was slightly more time-consuming but even more effective than the previous method. To establish protection in this area, the installation of metal hasps and padlocks was considered but, since these manhole covers were located on the public street, a tripping hazard would be created and undoubtedly result, sooner or later, in lawsuits.

Interviews with the electrical company management revealed the manhole covers were removed infrequently, seldom more often than annually, for maintenance purposes. A recommendation was therefore made to spot-weld the manhole covers in place. They were spot-welded in four places, the protection now established was effective, and attacks upon this type of transformer bank ceased almost immediately. The dissidents concentrated their efforts and ingenuity on other areas.

Barrier Clear Zones

The question of what distance establishes an adequate clear zone, both inside and outside the perimeter barrier, is quite controversial. Specific distances creating clear zones have been considered standard for some time. These distances are so great inside the protected area that, in most instances, it is uneconomical for the facility's management to agree to creating these clear zones. On the other hand, the majority of the perimeter barriers are established on the property line; therefore, control in creating any type of clear zone on the outside of the barrier is lost.

The type of material stored in outside storage areas and the proximity of the barrier to the buildings will usually determine the clear zone inside the barrier. As a rule of thumb, clear zones of at least eight feet should be created.

Where the barrier is within eight feet of a building, and the building could possibly be used to assist breaching the barrier, consideration should be given to increasing the height of the fence in the immediate vicinity and extending the additional height at least six feet on either side of the building.

If trees or utility poles are in close proximity to the barrier, efforts should be made to further deter attempts to breach the barrier through the use of these objects by fabricating and installing a metal "collar" around the tree or pole. The collar should be installed at approximately the same height as the fence. It should extend outward far enough so that scaling the pole at a greater height than the fence is not possible. These collars can be likened to the "rat guards" applied to hawsers when ships are tied dockside.

Obviously, the utility company must be consulted before installing this device on a utility pole. It may be necessary, in some instances, to construct the collar in two parts, securing it together with a metal hasp and padlock so it can be removed easily if servicemen must scale the pole.

Regardless of the size or depth of the clear zones, the area immediately surrounding the perimeter barrier must be kept clear of high weeds and other undergrowth. Such growth creates cover for intruders. It can also be used to secrete pilfered items, which can be placed near the perimeter barrier or thrown over it and recovered later. Growth-retarding or weed-killing chemicals should be considered for use in reducing this hazard.

Washouts and drainage ditches which breach the perimeter barrier, creating an opening of 96 square inches or larger, must also be dealt with. These hazards can usually be eliminated by driving metal stakes, approximately 8 inches apart, deeply into the ground or by installing barbed wire securely over the opening.

Piers, docks, and wharves, in some instances, will create still another security problem to contend with in establishing perimeter protection for the facility. These areas can usually be secured by fencing them *outside* the protected area and creating a sufficient number of vehicular and personnel gates. If this is not possible, the perimeter barrier should be brought to a point as close as possible to the piers, docks, or wharves; and gates should be used to secure the protuberance which actually projects into the water.

If it would be possible for an intruder using a small boat to approach the property, secreting himself and the boat under the wharf, it may be advisable to install barbed wire. This can most easily be accomplished at low tide. Wire should completely surround pilings of the dock or pier.

Frequently, construction for plant expansion or renovation occurs inside the protected area. This, obviously, will create an additional security problem. In some instances, the hazard created is great enough that additional security guards may have to be employed. In other instances, perhaps additional control will solve the problem. The amount of additional security required will be dictated by each specific instance. However, if the construction planned will cover a period of many months, and if it is physically possible, consideration may be given to installing a temporary barrier around the construction site and creating a temporary gate in the permanent perimeter barrier to facilitate movement in and out of the construction site. This, in effect, merely moves the original perimeter barrier some distance inside the protected area; the security established on this new barrier must be of the same degree as that provided by the remainder of the barrier.

Additional Barriers

In many instances, it may be necessary to construct an additional barrier inside the already protected area for the security of specific installations or equipment. This may require either a masonry or chain link type of barrier. The decision to further protect these installations or areas will again be dependent upon the criticality and vulnerability assessment previously made.

Some of the areas considered may be powerhouses, boiler rooms, water pump houses or water wells, gas-metering equipment, fuel storage areas, hazardous loads on railroad tank cars, or tank trucks and chemical storage areas.

This partial list of the installations or areas which may be considered includes those most commonly found at the industrial site. It may not, in many instances, be necessary to further protect any installations located inside the perimeter barrier. However, the study and analysis of the security plan must include these possibilities.

Attractive Nuisances

There are a number of installations commonly found at the industrial site which lie outside the perimeter barrier but still on company property, and which normally would not be considered in the perimeter protection plan because of the type of installation or its function in the facility's mission. These areas may include small lakes or ponds on the property, the facility's trash dump, escaping steam, or settling basins. These and other such installations or conditions are often considered attractive nuisances, particularly to children. They often result in injury to youngsters who might not have trespassed had a perimeter barrier been installed purely for deterrent and psychological reasons.

Additional perimeter protection might be installed in these areas, not necessarily for security reasons, but to reduce potential danger which could cause injury and death—and subsequent charges of negligence, resulting in lawsuits.

Where a body of water, or other natural obstruction such as a steep cliff or swampy area, forms a part of the perimeter, it in itself should not be considered an adequate perimeter barrier. Additional security measures must be provided for that portion of the perimeter.

The severity of the natural obstruction would dictate the additional protection necessary. It may require installation of a fence, or more frequent security guard patrols through the area, or possibly increasing the illumination of the protective lighting system in that particular area.

Once the perimeter barrier has been installed, the clear zones established, the security force deployed at the points of ingress and egress, inspections of the entire perimeter barrier on a regular basis are mandatory. These inspections should be conducted at least weekly, and their results submitted to the security director, who must initiate immediate remedial action to eliminate adverse conditions reported.

CHAPTER 4

Area Security

The perimeter barrier and its necessary protective aids is by itself considered only minimum security for the facility. Other ingredients must be added.

The next step in formulating the protection plan is to apply adequate security measures to outside areas of the facility. Formulating a security plan to adequately protect these outside areas will assist in developing the second line of protection.

In establishing this second line of protection it is important, first of all, that any storage in outside areas be organized and maintained so that it is as uniform as possible and does not present unnecessary obstruction to good observation.

Outside Storage

If material is stored in outside areas, or if products are stored outside the facility's buildings and these products are of reasonably high value, the stacks of this material should be arranged so as to make it difficult to remove items below the top of the stack. The items of higher value, wherever possible, should be those stored near the bottom of the stack.

Outside storage will normally be protected from the weather. If tarpaulins, or heavy plastic covers, are used, they can be secured by fastening the grommets to eyes imbedded in the cement or bituminous hardtop on which the items are stored. If locks will not be used, railroad seals should be used. This will assist the guard in conducting his tours through the area to immediately observe whether or not the stored material has been tampered with.

It is most important in establishing security in outside storage areas to insure that all material or products are stacked or piled neatly and uniformly. Aisles between stacks should be as wide as possible and free of any other debris. This affords the patrolling security guard good observation throughout storage areas. Later, it will be found that organized storage is mandatory if protective lighting is to be effective.

Housekeeping

Housekeeping in outside areas of the facility is as important as organizing storage. Rarely is an industrial facility found which is not guilty of having discarded machinery, unusable lumber, and, in general, plain junk thrown in a haphazard fashion in remote sections of the area being protected. These conditions create unnecessary hazards and must be eliminated. Additionally, when this condition exists, groundskeeping on the outside becomes more difficult and, invariably, during the growing season high weeds, grass, and brush cover the entire area where the discarded material is strewn.

All high weeds, brush, and grass must be removed before adequate security measures can be applied. All outside areas which are not paved or hardtopped should be seeded and maintained in almost the same condition as the lawns of a private home. The outside areas between buildings must also be kept clear to provide patrolling security guards clear observation between the buildings.

For security purposes, there are always objections to the presence of shrubbery or trees within the protected area of a facility, but these objections can somewhat be reduced if proper trimming and maintenance of the landscape are enforced. Obviously, trees and extremely dense growths of shrubbery provide cover for the intruder and do hinder proper observation by guards; therefore, whenever possible, this type of planting should be held to a minimum. The security director must recognize that modern approaches to management include providing pleasant surroundings for the employees. Management should not lose sight of the fact that reasonable attempts to landscape the grounds at a facility can be tolerated; however, landscaping quite often is overdone, creating problems which may require additional expense in installation of illumination and increased patrolling.

Outside Production Areas

At some factories, outside production areas are a part of the production facilities. These areas normally will not require any additional or special security measures while operations are in progress. This type of production area is usually restricted to such operations as forging, processing lumber, constructing pallets, and other such processes.

An exception would occur when valuable products or materials are being processed. If these areas are in close proximity to truck or rail shipment operations, additional fencing to protect them may be necessary.

Vehicles Parked in Outside Areas

It is not an uncommon practice for loaded or partially loaded trucks and trailers to be parked inside the perimeter barrier overnight or during the weekends; such a procedure is good security practice. However, additional protection should be afforded these loaded vehicles.

Truck trailers are normally equipped with loading doors at the rear, and some have additional loading doors on the sides. If the trailers are parked back to back and closely together, it is impossible to open the doors wide enough to remove items from the trailer bed. If a sufficient amount of trailers are normally being parked in outside areas, they should be parked close enough together so that it would be impossible to enter the trailer through the side doors or the rear doors without actually moving one of the trailers.

In all probability, trailers parked inside the protected area for any period of time will have the doors sealed. The doors should be further secured by padlocks, if the above procedure is impractical. The padlocks would be removed by the security guard when the truck departs the protected area. Security patrols through the truck park areas must include an inspection of the doors of these vehicles on every round.

Often facilities will have company-owned vehicles, such as station-wagons or small vans, which are used to perform local errands. These will remain parked within the protected area. Parking areas for these vehicles should be selected some distance from the production and manufacturing operations in an area that is well lighted. The vehicle should be completely secured by locking all doors.

These service vehicles should be parked inside the perimeter barrier, where they can be additionally protected by the area patrols. In many instances, it may be possible to park this type of vehicle in the vicinity of a fixed post where the security guard can constantly observe the vehicle.

The company stationwagon or van should not be authorized to park in any employee parking lots inside the perimeter barrier, or in any truck bays in either the receiving or shipping portion of the facility. If protection from the weather is to be given this vehicle, then arrangements should be made for the construction of a shelter or garage.

Employee Recreational Facilities

In many facilities, particularly those which are being constructed at the present time, areas have been set aside and developed to provide recreational facilities for the employees. Often these areas lie within the perimeter barrier and as often as not are immediately adjacent to the perimeter barrier.

In some instances, recreational facilities may be provided for employees outside the perimeter barrier yet still on company property. If these facilities are used by the employees during their lunch breaks provisions will have to be made to control the ingress and egress of employees between the recreational area and the main facility, establishing the same degree of control maintained at the regular employee entrance and exit.

Recreational areas which lie outside the perimeter barrier are often made available for use by individuals and organizations of the community. These conditions present a peculiar problem to the security director because of the public relations activities involved. The security force must handle breaches of security regulations or attempted breaches of these regulations with diplomacy but still maintain a firm professional attitude. When these situations exist, they may be tailormade for the individual or individuals who are planning unauthorized entry to the factory for whatever reason.

Unused Real Estate

For the past few years it has become customary for a company to purchase a tract of land considerably larger than that required to complete present construction plans, the unused acreage being intended to accommodate future expansion. When these unused areas exist inside the perimeter barrier and the land is normally not in public view, management is prone to do little toward maintenance. During the growing season, weeds, brush, and trees will appear and obscure observation almost completely. Any illegal activity which might be planned might well include use of these areas, since they cannot be observed by the patrolling security force.

In addition to reducing the effectiveness of area security, this undergrowth, during the nongrowing season, creates an unnecessary fire hazard if allowed to remain. The facility security director should use these two reasons to ensure that regular maintenance of an overgrown area is carried out.

Employee Parking Lots

Whenever possible, employee parking lots should not be established within the perimeter barrier. Parking lots should be established outside the perimeter barrier, if the configuration and the size of the property will permit. If the established perimeter barrier is already installed on the property boundaries and employees must park inside the barrier, the parking arrangements should be properly organized and additional fencing installed to enclose the parking area. The movement by the employees to and from their automobiles must be through gates to the parking area, and a security guard should be on duty at the gate.

Parking lots should be hard surfaced, and the parking plan arranged so that a maximum number of vehicles can be parked. Movement in and out of the parking stalls must be in a direction with the planned traffic flow. The simplest method of creating a planned, efficient parking lot would be to employ a local organization which specializes in this work.

In those areas of the parking lot where automobiles will be parked adjacent to the protective barrier, additional barriers constructed of substantial material should be erected some distance from the ground or installed directly on the asphalt surface, to prevent damage to the fence. Such barriers properly installed will also prevent a vehicle from being used to climb over the fence.

The parking lot plan should include reserved parking areas for such people as visitors, applicants for employment, or truck drivers who may be required to leave their cars in the parking lot for a number of days. These reserved areas should be created as near as possible to the fixed security guard posts controlling ingress and egress. This procedure will help guide people not familiar with the premises and will also provide additional protection for those vehicles remaining in the area for a considerable period of time.

Points of vehicular entry and exit through the protective barrier of the parking lot should be controlled by gates. If second- and third-shift operations are in progress, the gates should be closed and secured. They should be opened only during shift changes.

The security established for the protection of the employees' personal vehicles cannot be overemphasized. Unprotected parking lots create fertile grounds for car thieves, vandals, and automotive-parts thieves. Damage, loss, or theft of components from an employee's personal vehicle will affect morale.

Frequently, parking of personal vehicles belonging to executives, members of management, or office administrative employees is authorized within the protected area as a matter of convenience. If this cannot be eliminated, an examination of the area should be conducted to determine whether or not it can economically be fenced to restrict movement of production employees to and from or through it.

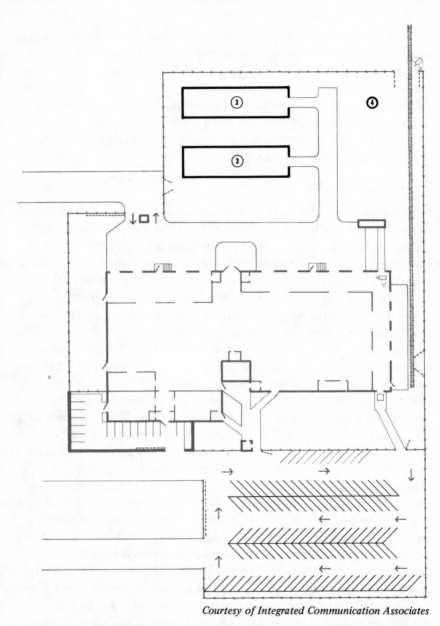

Courtesy of Integrated Communication Associates

Fig. 4-1. *Area security* must provide for security of (3) outside storage areas, (4) critical installations, in this case the water tower and employee parking lots.

Even though the vehicles may belong to trusted employees, they can be used on occasion by the pilferer, who will place items in the trunk through the use of a master key he has obtained. The items are recovered later, sometimes in the driveway of the owner's home late at night. In other instances, such vehicles have been used to transport stolen items out of the protected area; the item was taped to the chassis of the vehicle and recovered when the vehicle was outside the protected area.

Area Security Patrols

Unfortunately, the provision of supervised patrols for outside areas is often overlooked. Either foot or vehicular patrols, depending upon the size of the area involved, are undoubtedly the most important security measure that can be applied to outside areas.

Outside security patrols should be supervised by the installation of watchclock key stations at points that will insure that any route taken by the patrol will cover the maximum amount of the area. That maximum amount might be the distance physically covered, or the extent of observation possible without covering distance.

Outside security patrols, even more than patrols inside a building, must be conducted so as to insure that no pattern as to time or direction is established, because these patrols can usually be observed from areas outside the perimeter barrier. The frequency of the patrols will be determined only by the degree of security which need be established. In some instances, one portion of an outside area may be patrolled more often than others.

It must be emphasized that if there are outside storage areas, the patrols must include movement through these areas. If the protective lighting sufficiently illuminates the barrier and the area between the barrier and the guard's location, and if there are no objects or other means of concealment which could be used by an intruder who has successfully breached the perimeter barrier; the security guard need not move any great distance from the main buildings.

The facility site plan, Fig. 4-1, shows in heavy lines the arrangements for outside storage area security, critical installations, and the employees' parking lot.

CHAPTER 5

Protective Lighting System

It is difficult to place more importance on one particular aspect of a protection plan for an industrial facility than another; but without an adequate exterior protective lighting system, all those plans already formulated for the creation of an adequate perimeter barrier and establishment of area security would be most ineffective during the hours of darkness.

Consider for a moment a facility protected by a perimeter barrier which uses only that illumination installed for operational purposes as protective lighting. This condition frequently exists, and only after incidents occur and are brought to the attention of management is a protective lighting system installed.

The standards for establishing a protective lighting system at an industrial facility have been set by the Illuminating Engineering Society.[1] A good protective lighting system should direct adequate light upon bordering areas, glaring light in the eyes of an intended intruder, and little or no light on the security guard. There should be a high brightness contrast between the intruder and his background. This is accomplished by supplying an adequate amount of light and proper use of background material.

Industrial facilities differ widely in their physical character and size; therefore, it is difficult to attempt to list a specific set of rules which can be applied universally. What this chapter will provide

---

[1] *Practice for Protective Lighting. RP10-1956.* Illuminating Engineering Society, 345 East 47th Street, New York, N.Y. 10017. Also designated A.85.1–1956, and available from American National Standards Institute, 1430 Broadway, New York, N.Y. 10018.

is a guide to techniques which have been widely used and have undergone sufficient field testing to prove they are adequate.

The Effective Protective Lighting System

Either of two basic systems or a combination of both systems may be used to provide the protective lighting necessary.

The first method is to illuminate the perimeter boundaries and approaches to the boundaries; the second method is to light the area and the structures within the perimeter barrier.

If a protective lighting system is to be effective, it must accomplish the following purposes, and these provisions should be kept in mind during the entire analysis of the protective lighting system:

1. The system should be designed to insure that the illumination is sufficient to discourage and deter entry.
2. Should entry be accomplished, the intensity of the illumination should be sufficient to make detection certain.
3. Installation of individual lights should avoid throwing any glare in the eyes of security personnel or casting annoying lights into adjacent streets and highway traffic areas, neighboring facilities, or occupants of neighboring buildings. If glare can be directed at the intruder, it is most effective in handicapping him and reduces his ability to see the patrolling security guard.
4. The system must provide for complete reliability; a failure of any single light should not leave an intensely dark area in the system.
5. The system must provide special lighting for such locations as wooded areas, railroad sidings, water approaches, and points of ingress and egress on the perimeter barrier.
6. Adequate illumination must be provided. The amount of light required in certain areas depends upon the criticality of the installations in the area and their vulnerability to being successfully attacked or intruded on. The system must be engineered to eliminate heavily shadowed areas.
7. The system must provide means of convenient control and maintenance.

8. The system should provide for supplementary lighting that may include searchlights or other portable emergency lighting which is controllable and operable by the security force.

9. To provide maximum security, the system should have all poles and other equipment located inside the perimeter barrier, so that it is not readily accessible to damage or destruction.

Protective lighting should never be relied on alone but should be used with other security measures, such as patrolling guards, perimeter fences, and other physical protective devices. The facility site plan Fig. 5-1 shows the arrangement of an outside protective lighting system.

There are three primary methods of controlling the protective lighting system to turn on the illumination and turn it off at the proper times:

1. *Automatic controls.* These operate automatically by a photoelectric control installed on the protective lighting fixture itself, or in a central location with the control wired to several lights. The control is activated by the change in the amount of light passing through the photoelectric cell. The systems controlled by this device are automatically illuminated at darkness and extinguished at daylight.

2. *Timing device.* This system is controlled by a timing device which automatically engages a switch at a given time of day. The device can be likened to a timing device used in the home to turn appliances and lights in the home on and off at predetermined times. Care must be exercised when these devices are used to make sure necessary adjustments are made as daylight increases or decreases with the changing seasons of the year. Some of these devices are manufactured to automatically compensate for this difference.

3. *Manual operation.* The entire system or a series of lights within the system is turned on or off by a member of the security force or a member of company management. This is accomplished by means of a switch. However, human error and omission are involved in manual operation, and may result in the lights not being illuminated or extinguished at the desired times.

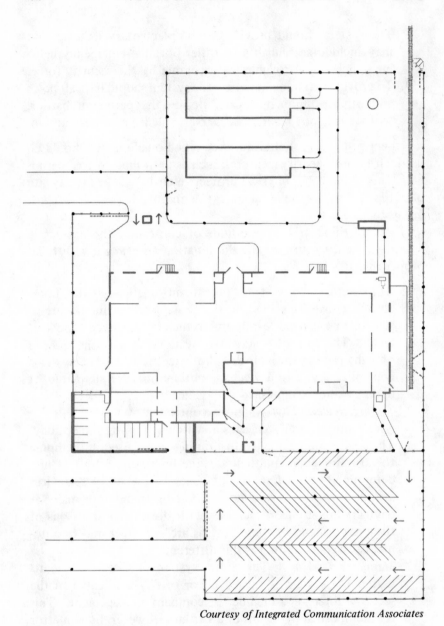

Courtesy of Integrated Communication Associates

Fig. 5-1. The outside *protective lighting system* starts with adequate illumination at the perimeter barrier to provide sufficient intensity throughout the protected area and insure elimination of heavily shadowed areas. Guards must be able to observe all outside areas during their patrols.

Types of Lamps

There are, primarily, four types of lamps or luminaires used in the protective lighting system. Each type of lamp and a general discussion of how each lamp operates appear below:

1. *Incandescent lamps.* These are the same type of light bulbs used in ordinary table and floor lamps in the home. They are the preferred type of luminaire for installation in the protective lighting system, because when the power is applied illumination is instantaneous.

2. *Mercury vapor lamps.* These luminaires produce a white light, directly or indirectly, which is caused by mercury and other gases acting with an electrical current. These lamps, although very popular today, are not as effective as the incandescent lamp, because they require a warm-up period initially, and in the event of a momentary power interruption the lamp will not re-light immediately.

3. *Sodium vapor lamps.* These are used occasionally in the protective lighting system. They produce a high intensity of light of a yellow color and may be of some value in a protective lighting system in areas where frequent heavy fog conditions exist. The yellow color has a more penetrating power than a white light, which is diffused during fog or falling snow.

4. *Quartz lamps.* These lamps emit a very bright, high-intensity white light and ignite almost as instantaneously as the incandescent bulb. The quartz lamp is suited to very high wattage, normally over 1,000 watts per lamp. In the protective lighting system, it is not uncommon to find quartz lamps of 1,500 and 2,000 watts. These lamps can be effectively used to illuminate the perimeter barrier or any other outside area and are particularly suited to increasing the illumination at critical and vulnerable installations or areas.

Lights specially engineered for outdoor use and offering high illumination and durability are excellent selections for the protective lighting system. An example is shown in Fig. 5-2.

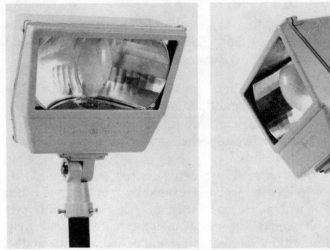

Courtesy of the Lamp Division, General Electric Co.

Fig. 5-2. Floodlight luminaires enclosed in weatherproof housings mounted on poles or on building walls are directed to illuminate the perimeter barrier, storage areas, or outside open terrain.

Seeing at Night

Regardless of the type of luminaire or lamp which is used in the protective lighting system, several conditions must be met if activity in the areas under surveillance is to be quickly detected and the action readily identified.

Brightness is an essential factor. The security guard patrolling in the area between the building and the perimeter barrier, and between the buildings of the facility, is required to make observations at long distances. In addition, he must be able to distinguish low contrast objects. He must be able to quickly observe an intruder who may be exposed for only a matter of a few seconds. These vision factors are improved by higher levels of brightness.

In predominantly dark, dirty surfaces in heavy industrial areas, more light is needed to produce the same brightness as around plants having outside areas covered with light-colored gravel or cement. Areas around plants which contain a preponderance of shrubbery, grass, lawns, and large landscaped areas would also require a higher level of brightness, because light is absorbed by growth which also creates many concealing shadows.

Contrast is another requirement. It is obtained when the same light falls on the subject that a guard is looking at and on the background behind the subject, and the guard must depend on the difference in the amount of light being reflected. As an example, if an individual with dark clothing is standing near a wall that has been treated with whitewash, the contrast between the individual's dark clothing and the white wall would be easily discernible. Just the opposite would occur if the individual were wearing white clothing, standing near a whitewashed wall, and far enough away so that no shadow was cast. This condition makes it almost impossible for the guard to observe the intruder, particularly if at the moment of observation the intruder is not moving.

When the intruder is darker than his background, a guard will see primarily the outline or *silhouette* of the individual. Intruders who depend upon dark clothing and even darkened faces and hands could be foiled if light finishes were used on all the lower parts of walls and structures. An intruder who is silhouetted by the protective lighting is more easily observed when he is moving than if he remains still. When observing silhouettes, the guard must keep in mind the conformation of the human body while the individual is standing, crouching, or bending over.

Elimination of shadows is vital. Dark areas created by shadows of illuminated objects are the greatest ally of the intruder, who will habitually move from shadow to shadow. These shadows create hiding places and, if they are sufficient, make it fairly easy to ambush the guard. Shadowed areas in both indoor and outdoor lighting must be eliminated. Therefore the protective lighting system should be designed to eliminate as much as possible, heavily shadowed areas. The beams of light of all protective lighting fixtures should overlap, so that if one lamp burns out the portion of the area illuminated by this lamp is not plunged into complete darkness.

Unorganized outside storage areas are almost impossible to effectively illuminate so that all shadowed areas are eliminated. If the storage is organized, the majority of the shadowed areas can be eliminated or reduced in size so as to be ineffective for use by an intruder.

Types of Systems

Generally speaking, there are four types of protective lighting systems or methods. One or more types may be used in a single system at a facility; that is, within the system as a whole, four subsystems may be considered.

1. *Continuous lighting.* This is the most common type of protective lighting system. It consists of a series of fixed light fixtures designed to provide a continuous flood of light to a given area. The beams of light should always overlap.
2. *Standby lighting.* Standby lighting differs from the continuous lighting system in that it is not used in the protective lighting plan on a regularly scheduled basis. The light fixtures are not continually lighted; they can be either automatically or manually turned on when needed by the security force. Standby lighting may be defined as those lights which are turned on when an anti-intrusion alarm system is activated. As a simple explanation of this type of lighting, the light in the refrigerator may be termed standby lighting, for it is illuminated only when the door of the refrigerator is opened.
3. *Movable lighting.* This subsystem of lights consists of manually operated, movable searchlights or floodlights which are used to supplement the continuous or standby lighting systems as a particular emergency situation may dictate. The movable lighting may be on fixed mounts, or it may be manually moved or moved on motor-driven vehicles.
4. *Emergency lighting.* Emergency lighting is used during times of a power failure or other emergency situation when the normal lighting system is out of operation.

Auxiliary Power Sources

If the protective lighting plan is to be considered entirely reliable and efficient, an auxiliary power source should be available for use in the event the primary power source fails. Such power sources may be provided by storage batteries or a generator driven by a suitable internal combustion engine. The auxiliary power source should provide for the automatic takeover to transfer the lighting system from the normal utility source to the duplicate source, if the utility supply fails.

Perimeter Barrier Illumination

This discussion on the illumination of the perimeter barrier will be confined to nonisolated fence boundaries and semi-isolated fence boundaries.

A nonisolated fence boundary consists of fence lines where the protected property is immediately adjacent to operating areas of other facilities, or to public thoroughfares where nonemployees may move about freely, and legally approach areas in close proximity to the barrier. The nonisolated fence boundary differs from the semi-isolated fence boundary in that the width of the lighted strip inside the plant property is increased, and the width of the lighted strip outside the plant property is decreased.

Semi-isolated fence boundaries are illuminated by directing the cone of light outward from the barrier, so that security guards patrolling in back of the pole lines would remain in comparative darkness and a would-be intruder 150 to 200 feet outside the fence would be visible to the guards. If the street-light type luminaires are used, no objectionable glare will result on the property outside the perimeter barrier.

The illumination of nonisolated and semi-isolated fenced boundaries is obtained by spacing the lights approximately 150 feet apart and at a height of approximately 40 feet. In both instances, the location of the light should be approximately 30 feet inside the perimeter boundary. In the interests of economy, if building walls are within the specified distance from the perimeter barrier, luminaires can be installed directly on the face of the building at the prescribed height. It may be necessary, in some instances, to extend the luminaire from the building to place the light source within the prescribed distance from the barrier to be lighted. A quick estimate of both costs and a study of the activity between buildings near the perimeter barrier will determine whether or not individual poles should be installed or if the fixtures could be mounted on the buildings.

Illumination of Perimeter Entrances

The illumination of points of ingress and egress on the perimeter barrier are normally gates. However, in those instances where the walls of a building form a portion of the perimeter, these entrances would be doors or other passages. Additional lighting at these points must be considered whether or not they are authorized for constant or occasional use after the hours of darkness.

Active pedestrian entrances should be provided with two or more lights which furnish adequate illumination for recognition of persons and examination of credentials.

Vehicular entrances must have two or more luminaires located so as to facilitate complete inspection of passenger cars, trucks, and their contents. It may be desirable to extend the illumination at these points of conveyance to insure the space underneath the vehicles is adequately illuminated. This procedure is normally used at railroad gates to provide the security guard with sufficient illumination for an inspection of the undersides of freight and tank cars.

Inactive entrances,—entrances which are normally locked and infrequently used during the hours of darkness—should be equipped with additional illumination. Provisions should be made for auxiliary switching so that these points can be additionally lighted.

Building-Face Boundaries

Building-face boundaries consist of the faces of those buildings which are on or within 20 feet of the perimeter barrier and where the public may approach. In these cases, fences may or may not be installed and the security guards may be stationed or patrol either on the inside or the outside of the buildings. Doorways and other insets in the building face should be illuminated by installing a luminaire directly over the opening to eliminate shadowed areas created by other illumination not under the control of the facility's management.

Waterfront Boundaries

Waterfront boundaries consist of property lines, whether fenced or unfenced, at or adjacent to bodies of water. The lighting system should conform, basically, to fence barrier standards, for perimeter barriers, except in the case of non-navigable water of less than 15 feet in width. Such narrow water should be illuminated in a similar manner to that prescribed for nonisolated or semi-isolated fenced boundaries and handled as though the body of water were not present. There should be no concealing shadowed areas on or near the water or close to a seawall or bank.

The United States Coast Guard must be consulted for approval of proposed protective lighting adjacent to navigable waterways.

Piers and Docks

Where piers and docks form a part of the industrial complex, the installations must be safeguarded by illuminating both the water and land approaches. The land approaches to an industrial pier or dock are generally vehicular streets or open areas, and the lighting fixtures should be the same as those used for lighting the remainder of the industrial thoroughfares.

The pier or dock area is often furnished supplementary lighting capable of being directed by the security guards as the need arises. Caution must be exercised to insure that lighting does not, in any way, violate marine rules and regulations and causes no glare to affect waterway pilots. Again, the United States Coast Guard should be consulted for approval of the proposed protective lighting adjacent to navigable waterways.

Depending upon the criticality of the operations carried out at piers and docks, it may be advisable to install protective lighting under the docks. These lights will have to be enclosed in waterproof housings.

Areas and Structures Inside the Protected Area

The areas discussed below lie within the perimeter barrier and consist of industrial thoroughfares, yards, storage areas, and unoccupied or occupied working areas. In some instances, buildings under construction may need to be included.

Thoroughfares over which heavy truck traffic and primary industrial movements take place are usually lighted by means of suspension luminaires or floodlights, depending upon the problems involved. Main trucking roads which are some distance from buildings may be illuminated in a manner similar to those standards set forth for the illumination of open yards.

Open yards consist of unoccupied land only, and a low level of illumination will furnish the protective lighting requirements in these areas. The standards mentioned earlier specify a minimum light level, which consists of floodlights mounted on poles installed throughout the area.

The illumination of open yards in areas adjacent to the perimeter barrier should be in accordance with the standards, mentioned earlier, which apply to the illumination of nonisolated and semi-isolated fenced boundaries. The illumination of all open yards within the protected area, except those described above, should be uniformly lighted, with sufficient illumination on the ground to assure ready detection of persons by the security force.

Outdoor storage areas include material storage areas, railroad sidings, and parking areas. The lighting units must be placed so as to provide an adequate distribution of light to illuminate aisles, passageways, and recesses, the object being to eliminate shadowed areas where unauthorized persons may seek concealment.

The illumination of *vital structures* and buildings under construction within the protected area should conform to the following standards:

1. The vertical surfaces of all vital structures which could be easily and seriously damaged at close range must be lighted to a minimum height of eight feet.

2. Vital structures which could be easily damaged from a distance with serious effects on production should not be lighted. This lessens their visibility to attacks from outside the perimeter barrier.

3. The protective illumination of areas occupied by construction projects should be approximately the same intensity as those standards established for the illumination of open yards.

The protective lighting used to safeguard *shipping and receiving areas* is usually furnished by that lighting installed to support operations. In rare instances it is necessary to increase the illumination of the operational shipping and receiving truck docks. See Figs. 5-3 and 5-4.

Courtesy of Phoenix Docklite

Fig. 5-3. Illumination of docks to support operational requirements when properly applied, also supports the security program.

In examining the intensity of illumination at the docks, emphasis should be placed on adequate illumination in three areas:

1. Sufficient illumination must exist on open docks—docks constructed outside the peripheral walls of the building—to eliminate all shadowed areas and provide the patrolling security guard with excellent observation of the entire dock area.

Courtesy of Phoenix Docklite

Fig. 5-4. Use of protective illumination to light the interior of a truck trailer. The well-lighted interior deters pilferage.

2. The illumination below the floor level of the dock is usually installed about midpoint between ground level and the dock floor. It should be sufficient to insure that ground level areas between and under parked trucks can be readily observed. Lighting must be sufficient to detect even small packages which may have been inadvertently or purposely dropped from the dock floor for possible recovery by a driver.

3. Sufficient illumination of the interior of trucks being loaded at night must be furnished not only for operational and safety reasons but to insure that security guards assigned patrol duties on the shipping or receiving docks can readily detect surreptitious activity by dock personnel as they are working within the confines of the truck or trailer.

Lighting Equipment

There is a wide range of lighting equipment available, a great deal of which find some application in the protective lighting system, but none of which is universally adaptable for all purposes. Broadly speaking, there are only four general forms of lighting units suitable for, and in use in, developing a protective lighting system. These four classifications are: (1) floodlights; (2) street lights; (3) Fresnel units; and (4) searchlights.

Floodlights are designed to form the light into a beam so that it can be projected to distant points or for illuminating definite areas. Floodlights are specified in wattage and beam width. Beam width is expressed in degrees and defines the angle included by the beam. To roughly classify floodlights according to beam width is to use the terms narrow, medium, and wide. Floodlights may also be open or enclosed; the enclosed fixture is equipped with a cover or door glass to exclude rain and dust. The enclosed floodlight is preferred, since it protects the lamp and reflecting surfaces from damage due to weather and any deteriorating effect of exposure to the atmosphere.

The floodlight may be successfully used to illuminate boundaries, fences, areas, buildings, and for special lighting of vital structures or areas.

Street lights are rated by the size of the lamp which the fixture accommodates and the characteristics of the light distribution. It is quite a common practice to use up to 500-watt lamps in multiple street-light luminaires.

The distribution of the light may be symmetrical or asymmetrical. In a symmetrical distribution, the distribution of light is approximately the same in any vertical plane which passes through the axis of the luminaire. This kind of light distribution is in wide use in illuminating large areas, where the luminaires are located centrally in the area being lighted. Street lights are also very effective to illuminate entrances, exits, and special boundary conditions.

Asymmetrical distribution may take one of many forms and makes a higher utilization of the light by directing it by reflection, refraction, or both into the direction needed. An example of the use of the asymmetric distribution is for the illumination of boundaries where the fixture is located inside the property and the light is delivered largely outside the fence. The asymmetrical distribution is also used to illuminate roadways where the pole, of necessity, must be placed some distance from the roadway.

The *Fresnel lens luminaire* used in the protective lighting system delivers a fan-like beam of light approximately 180 degrees in the horizontal and from 15 to 30 degrees in the vertical. It is used to increase the protection to the facility by directing the light outward to illuminate the approaches and inflict glare on the would-be intruder while the security guard remains concealed in comparative darkness. The 300-watt lamp is commonly used with this unit.

The application of Fresnel or glare projection lighting is limited to locations where advantages may be taken of its unique characteristics without causing an objectionable glare to neighboring activities. An intruder approaching an area protected by Fresnel units is faced by, and his observation limited by, a glare of light similar to that of an individual attempting to observe an approaching automobile with the headlights lighted.

The glare projection method of protective illumination is suitable for lighting water approaches to the protected area. Caution must be exercised, however, and approval must be obtained from the United States Coast Guard if these waterways are navigable as already mentioned.

When *searchlights* are employed in the protective lighting system, they are usually of the incandescent type because of the small amount of maintenance required for their operation as well as their simplicity and dependability.

Searchlights are rated by diameter of reflector and lamp wattage. Their usual application in the protective lighting system falls within the range of 12 to 24 inches in diameter, using lamps of 250 to 3,000 watts. The beam angles of these searchlights range between 3 and 8 degrees.

The mountings for searchlights are normally controlled manually and are of the trunnion and pedestal types. This type of searchlight may serve in the protective lighting system if it is mounted on top of guardhouses or in other places where its control is in the hands of the security force.

The primary use of the searchlight in the protective lighting system is to permit exploration of areas within or outside the property, or to supplement, when and where required, the fixed lighting within the property.

CHAPTER **6**

Building Security

Protection provided for the buildings comprising the industrial complex must be tailored to meet the security requirements of each individual building and specific areas within that building.

It should again be emphasized that to establish the degree of security required for a specific building, buildings, or areas within these buildings, an intelligent assessment of the criticality and the vulnerability of the separate structures and the areas or processes within these structures must be accomplished. Without such assessment and the establishment of priorities, a successful protection plan is difficult to formulate.

The establishment of physical security for the buildings inside the protected area provides further lines of protection and thus strengthens the overall protection of the facility.

In establishing the security for a single or a number of buildings, all parts of each building must be considered. A logical sequence in the examination of each structure entails starting this study at the peripheral walls and moving progressively through each function within that building. It will simplify matters if the examination of the building internally proceeds in the direction of the production flow, from the areas of input of the raw material, through the production processes, and finally to the areas housing the finished material, or product. Some of the important factors that should be taken into account are considered graphically in Fig. 6-1.

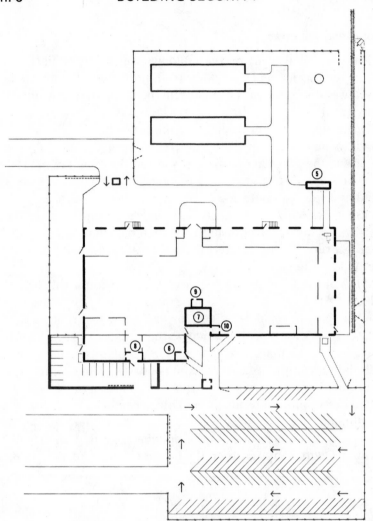

Courtesy of Integrated Communication Associates

Fig. 6-1. Planned building security includes protection of storage and refueling buildings (5), and transformer areas (7). Inside the buildings, security must be tailored to furnish the degree of protection required and may involve relocation of the personnel office (6), first-aid room (9), or the company sales store (10). The main lobby (8), where the switchboard is located and where visitors are controlled, requires careful study.

Doors and Windows

The peripheral walls of the buildings which form a part of the perimeter barrier and the additional protection required where windows, doors, or passages breach the walls have already been considered.

The examination now should confine itself to a study of the windows and doors on the peripheral walls of the building which are inside the perimeter barrier. If window openings are in production or warehousing areas and the areas immediately outside the windows are in remote locations, it may be necessary to further secure these windows by installing screens or gratings over them. This will eliminate the possibility of material being thrown out for later recovery. Perhaps merely locking the window will prevent this method of pilferage.

Material used for protection over these openings will be determined by the type of processes occurring immediately inside. As an example, if small television or radio tubes are being stored adjacent to this type of window, the installation of chain link fencing with openings two inches square would probably not furnish the protection needed; screening material of smaller mesh may be required.

If these windows do not present this type of hazard, it is quite possible no other measures need be initiated. If the windows open directly onto an area outside the building, one that is, or will be, furnished with more security than the remainder of the building, it is quite possible that no further treatment of these openings is necessary.

The doors breaching the peripheral walls of the building and the walls between departments, operations, offices, production areas, and other restricted areas must be examined on an individual basis. This examination will determine the desired door construction and the type of locking device which should be installed.

To arrive at an overall protection plan for each individual building which will be commensurate with the degree of protection required yet not impede the orderly flow of personnel on a routine or emergency basis nor reduce operational efficiency, each such opening will require individual treatment; security must be tailored to each specific opening. Whether or not any special security measures may be required at any specific door will depend solely upon the degree of security required to protect functions of the plant or products stored within the area.

After having examined all openings in the peripheral walls, to follow a logical sequence start at the employee entrances.

Employee entrances are those doors which are authorized as points of ingress and egress for employees. They may be manned by a security guard at all times or merely during shift changes, depending upon each local situation—providing, of course, that employee control is not established at another point on the perimeter barrier such as at the parking lot gate. Once the employees have entered, the door is no longer needed until the shift departs, and it should be secured.

Employee entrance doors must always be considered as emergency exits. Should an emergency arise, regardless of any other designated emergency exits, some employees may be in a state of near panic and may attempt to exit through the door they are most familiar with. Therefore, these doors should be secured with a local alarm locking device to insure that they are not used surreptitiously and yet to permit rapid egress in the event an emergency does arise.

Emergency exit doors should receive the next priority in the security study. Since these doors are not intended for operational purposes, they should be secured with at least a local type of alarm and panic hardware. Local alarms will normally create a deterrent effective enough so that employees will not be tempted to use these doors except during emergencies. Since emergency exit doors are not intended for operational use, further security can be established by removing door handles, door knobs, or even any type of locking device which may be installed on the outside of the doors.

Fire doors are installed to close automatically should a fire occur. Under normal conditions, these fire doors will remain open. They may be designed for operational use as a secondary function.

Fire doors are most often found in dividing or fire walls between departments or production areas, and shipping and receiving areas.

If it is desirable to close and secure these doors as a security measure during specific times of the day, there is no objection to using these doors for this dual purpose. Should a fire occur, the door would automatically close, retarding the spread of fire, and no fire safety regulation or life safety regulation will be violated.

Dock doors consist of overhead and personnel type doors. Many overhead doors are manufactured with locking bars. A device to operate these bars can be operated from inside the building merely by turning a handle. Overhead doors must be secured in their lowered, or closed, position with a key-operated locking device. This can be accomplished by either drilling holes through the locking bars or through the track, just above the lowest door roller, and using a padlock by inserting the shackle through the hole.

Personnel-type doors in the dock area which are used by truck drivers must be controlled, either by the clerk assigned to the docks or by a supervisor. These doors can be adequately secured and control established by the use of an electrical strike operated from a remote location within the building. The person desiring entry into the dock area makes his presence known, and the individuals authorized to permit entry establish his identity. He is then admitted by activating the electrical strike which releases the security device installed on the doors. Examples of electrical strike units are shown in Fig. 6-2.

The electrical strike, more often than not, is activated merely by depressing a button remotely located from the door. Under some circumstances, it may be wise to install a key-operated electrical switch rather than the button type.

Doors separating warehousing and production areas can either be operational doors or, as previously discussed, fire doors. The doors normally remain open while warehousing and production are in progress. However, during the second and third shifts, warehousing operations may be closed down. To further secure the warehousing operations, doors between the departments should be closed and secured.

Again, if fire doors in this area are being used for operational purposes, there is no objection to closing the fire doors between these two departments and securing them in the closed position. Be certain this does not block an emergency route to a designated emergency exit door intended for use by the personnel in the production department.

 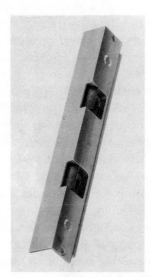

Courtesy of Folger Adam Co.

Fig. 6-2. The single (Left), and double (Right), electrical strikes permit doors to remain secured and to be opened from remote locations after an identification is verified.

The design of the building and the number and location of fire exits usually eliminate the need for employees in the production department to exit through either the receiving or shipping warehousing department. If they must exit this way, it may be necessary to construct a personnel door which breaches the wall. This should be designated as an emergency exit only and secured by those methods outlined previously.

Storage, supply, and toolroom doors should remain closed and secured unless operations are actually in progress. Supply and toolroom doors which are used continually during operating shifts may be fitted with "Dutch" type doors: doors where the lower half can be secured and used as a work counter, while the upper half remains open. This installation will greatly assist in restricting unauthorized access to these rooms. Remember, the locking devices on each half of the door must offer equal security when operations have been terminated and the area is totally secured.

Supply and toolroom areas are often constructed with industrial chain link fencing. These areas will often be built some distance from any peripheral or internal walls of the building: that is, the area forms a type of cage. When these areas are constructed, the sides of the enclosure must either be anchored to the existing ceiling or enclosed at the top with the same type of material used in the construction of the sides.

The doors to storage areas which contain no restricted flammable or explosive material should be constructed of solid wood; there is little requirement that a window be installed in this type of door. The overall construction of the door should be solid wood rather than two pieces of plywood installed over a frame.

Doors in hot working areas, which are left open for ventilation purposes and not intended as emergency exits or operational doors, can be secured by installing chain link fence doors or chain mesh doors over the opening to reduce or eliminate unauthorized traffic. Doors which lead from within such areas as boiler rooms, steam-generating areas, or from areas in which hot metal or other hot items are being processed, and which are also intended as emergency exits, will require that additional protection be installed to insure they can be opened and rapid egress effected. This is of utmost importance, because the hot working area is especially prone to fire emergencies.

Quite often the areas immediately outside the facility's boiler rooms are used as light-storage areas. These same areas are often used as temporary work areas where pipes are being cut, threaded, and fitted. If the open boiler-room door creates a security hazard but is required operationally, it may be necessary to install additional chain link fencing to enclose the work area immediately outside the boiler room. In this event, provisions may be made to insure that the personnel type gate installed on the chain link enclosure can be readily opened, should an emergency require evacuation of the boiler room or the enclosed area.

Doors between working and idle departments may have to be secured. In some instances, as an example, during second- and third-shift operations where production is somewhat curtailed, it may be necessary to secure doors between working and idle departments to reduce the possibility of employees entering areas in which they are not authorized.

When these passageways are not needed for emergency egress, effective protection is established by installing locks with a keyway on both sides. Merely to install hand-operated locking bolts or bars on the idle department side of the door does not furnish the degree of security which is needed, particularly if the requirement dictates that this type of door must be secured.

Doors between production and office areas should also be inspected. Almost every industrial facility has plant offices located in one of the production buildings. Routine business requires that line supervision and members of management move between the office areas and production areas.

Little or no security is normally required during the period that the office areas and production areas are both in operation. Usually a restriction placed upon production employees suffices to insure that unauthorized movement through the opening does not occur. However, when second- and third-shift operations are in progress, and the office areas are not occupied by the entire administrative staff, the doors separating these two areas should be secured with a key-operated locking device having keyways on both sides of the door. This would facilitate movement of line supervision into the office areas, if it were required and they were authorized to do so.

Doors or gates to danger areas, installed as the peripheral protection for such areas as high-voltage transformers or storage areas for explosive or flammable materials, should be constructed of solid wood and metal or solid metal, depending upon the location of the area, the type of material stored in the area, and whether or not an explosion or blow-out wall exists.

However, these doors must remain secured at all times and keys to operate the locking devices installed on these doors should be restricted and issued only to those individuals who need to and will be authorized to enter the protected area.

Doors to restricted areas, such as telephone equipment rooms, research and development operations, and such operations as quality control should be equipped with two locking devices, a dead-bolt type and a latch type. The door should also be equipped with an automatic pneumatic door closer.

Individuals whose ingress and egress is authorized should understand that these doors must remain closed and secured at all times. The installation of an automatic door closer will normally provide this protection. The locking device with a spring-operated latch should be activated and used during operational periods. This would eliminate having to use a key to get out of the area. A button-type cypher locking device installed on this type door would totally eliminate the need for a key.

The second locking device, referred to as the "security lock," should have a keyway both inside and outside the door. This locking device is activated during the periods when operations within the area have been discontinued.

Doors to specially restricted areas where important or vital classified company proprietary information, government classified information processes, or hardware are stored, require special treatment.

The requirements for securing areas involving government classified information are delineated quite clearly in the Department of Defense manual, *Industrial Security Manual for Safeguarding Government Classified Information*, also available from the U.S. Government Printing Office. If this condition exists at a facility, the responsible government official will already have established the security requirements.

When the restricted matter stored in these areas pertains only to company information, material, or projects, the classification must be established before the degree of security required can be ascertained. A good rule of thumb is that areas containing vital company proprietary information should receive protection of no lesser degree than that required by government standards of like security classifications.

Doors to "hot item" storage areas where highly pilferable material or material of high value, small enough to be easily pilfered, is stored, should receive special treatment. These hot item rooms or storage areas may include "OS&D" (over, short, and damaged) material.

These rooms or areas could be located anywhere within a facility but would normally be found either in the shipping warehouse or shipping dock area. The items may be secured within rooms protected by masonry walls, chain link fence, or, on occasion, fencing or even lighter materials. The degree of security established at these doors is directly dependent upon the items or material stored within.

Vault doors, or heavy metal doors installed for the protection of precious-metal storage rooms, monies- and record-storage areas, are usually of heavy fireproof construction. They are secured with a three-way combination lock rather than a key-operated device in order to afford adequate protection.

Vault security must also include control of the combination to the lock. Combinations must be changed immediately when individuals with knowledge of the combination no longer require it or leave the facility permanently.

Door-locking schedules and key control are important factors in the overall building security plan. Locking schedules must be established in minute detail and accomplished at specific times to insure that internal security is being maintained. They must be constantly reviewed to insure that supervisors or others in charge of areas or operations have not arbitrarily changed the schedule which is being supervised by the security force.

Often temporary changes in operational patterns will require that doors remain open longer than prescribed in the security-force schedule. Temporary changes in the schedule must be disseminated to the security force. When operations return to normal, the temporary locking schedules must again be changed. Often, under these circumstances, the areas or operations are placed at risk merely because of a breakdown in communication between operating personnel and the security force.

When establishing the locking schedule, it is also important to determine whether or not it is necessary that doors be checked by a second person after they have been secured. When this procedure is necessary, the security force should be charged with the responsibility. They must be as sure to check the security of these doors as they are to discharge their responsibilities of securing those that they are charged with actually locking.

Security of Special Areas within Buildings

The areas discussed below have been chosen because most industrial facilities, regardless of their size, will include them in one form or another.

Cafeterias or lunchrooms will normally operate on a schedule that permits employees to use this service from the morning coffee break through lunch and also the afternoon coffee break. Some cafeterias remain open during the second and third shifts, if a sufficient number of employees are employed to support these operations. However, the majority of the cafeterias and lunch areas are closed and secured at the end of the first shift.

Whether or not the cafeteria is concession-operated, money-handling operations should be examined, and refrigerators used to store food should be secured. If the area contains vending machines available to employees, arrangements should include provision for closing off and securing the kitchen and storage areas separate from the lunchroom and vending machine areas.

Credit unions will not normally handle cash in any quantities, since most of the transactions are by check, but the area must still be considered for additional security and most surely must be included in the building locking schedule.

Even though individuals assigned to operating the credit unions are not always company employees but people who are paid by the credit union, plant management has a moral obligation to insure that these people are properly identified and controlled and that the records are properly safeguarded.

If the credit union location is not within the protected area and does not occupy company-owned facilities, it would normally be disregarded by the security director.

Company stores are areas in some manufacturing facilities where company products are sold to employees on the premises. The examination of the company store activities and procedures should include analysis of the procedure used to supply the company-store stock and the method of accounting for these items.

If cash transactions are involved, depending upon the amount of cash on hand on any particular day, provision must be made to insure safe and proper handling.

Company-store hours should coincide with the end of shifts and employees purchasing items in the company store should not be authorized to return to their work areas or locker rooms with these items. Procedures must be established to insure the items are under control, and bags or cartons properly secured and identified with cash-register or hand-drawn receipts.

The employee should be required to immediately depart from the building after a purchase has been completed or surrender the package for storage until the end of his shift.

Employees' locker rooms must always be included in the guard tours, not necessarily because of the theft of any company property therein, but because they are often used to store material pilfered from the production and warehousing areas. The material is stored in lockers pending an opportunity to remove it from the facility. Additionally, situations will always occur when employees steal from one another, creating a distinct problem of morale.

Regardless of the method used to secure the individual lockers, management should retain a key so that unscheduled inspections of the lockers can be conducted. The security force should never be authorized to inspect the insides of lockers unless a member of management is present.

Quality control areas are important, even though quality control measures and procedures may not be considered proprietary information and the records maintained in this department may also not be considered proprietary. The area should be secured against entry after normal operations have ceased, particularly if the finished products used in testing are still available.

Even if management attaches no importance to the records pertaining to quality control, they should be considered by the security director as being proprietary and should at least be secured in locked security containers.

Pilot operations should be located some distance away from the normal employee traffic pattern whenever possible, and they should be secured against unauthorized entry and unauthorized observation by those passing in the vicinity of the operations.

Pilot operations may be secured by erecting an economical barrier using plywood or chain link fence with tarpaulins secured to the fence from the inside. Doors into the area should be controlled. Again, the amount of security applied to these operations will depend upon the classification assigned to the project.

Display areas for finished products, when they are open to company and outside salesmen, must be secured when not actually in operation. Control should be exercised to insure that items on display are not being pilfered when operations are in progress.

The degree of control at the entrances and exits to display areas will depend upon the type and size of the items on display, the category of individual who will be authorized to enter the area, and the number of these individuals involved.

In some instances, as in furniture manufacturing, these areas may be secured for a matter of several weeks prior to the showing of new models. When this condition exists, the security applied to the display area must be sufficient not only to insure that the items cannot be physically removed but also that they will not be observed by unauthorized persons.

There are few industrial plants which do not have *plant offices* located throughout the production and warehousing operations. These offices are normally manned only during the times that shifts are actually in operation. In some instances, offices may be manned only during the first shift, although second shifts and third shifts are in operation in the manufacturing areas. These offices are normally constructed with wide expanses of glass on all sides to facilitate observation of the areas of responsibility of the occupants.

Doors to these areas should remain secured to deter surreptitious entry by unauthorized employees, but, more importantly, production data, time schedules, blueprints or other such material should be kept in locked file cabinets or locked desks when not in use.

Relocation of Activities

Often it is desirable, and sometimes necessary, to improve security by relocating the activities of certain departments or functions. These activities were often established without regard to internal security, but since their establishment physical changes have taken place. The result now is that security hazards are present which can be eliminated or reduced by relocating the facility elsewhere.

During the physical examination of the facility, if it is determined that the overall protection plan or the protection plan applied to a specific area could be made more effective by relocating a specific facility, the security director can justify his recommendation to management by presenting all the pros and cons relative to its relocation.

There are primarily five activities which often need to be relocated: (1) guardhouses; (2) personnel offices; (3) company stores; (4) display areas; and (5) first-aid rooms.

Guardhouses, more often than any other activity, should be closely examined for possible relocation. Many times these structures were correctly located originally but are presently ineffective and inefficient because of physical changes which have been made at the facility.

When the overall security plan has been formulated, guardhouse sites should be reevaluated to insure that guards manning these points can operate and perform their functions at peak proficiency.

When personnel offices are not located immediately adjacent to, or do not themselves have, peripheral entrances, consideration should be given to relocation. Ideally, personnel offices should be located so that entry can be accomplished from outside the protected area. If this is not possible, routes from the building entrance to the personnel office should be as short as possible, and controls should be established so the applicants cannot inadvertently or surreptitiously move into office, production, or warehouse areas.

If, the personnel office is not at such a location it may be desirable to relocate it, and preferably at a position in a direct line of observation from the manned security control point where the applicant for employment will be admitted.

The company store can often be relocated to increase the security of its operations, particularly if sales are being made to nonemployees. Often families of employees are authorized to trade in the company store; or a company store may double as an outlet for finished products which are made available to the community as a whole. In this situation, those considerations given to relocating the personnel office should be applied to the company store, primarily for the same security reasons.

Display or sales areas may also be examined to insure that the area in which they are located is not only the most secure, but that the location would not require outsiders to traverse any areas in which they may become injured or would give them the opportunity to otherwise violate security. When these areas are open to buyers or salesmen not employed by the company, it is much more important to shorten the routes from the outside and to establish more restrictive controls.

First-aid rooms located in the plant administrative office areas will often create a security hazard which could be eliminated if the facilities were moved into the manufacturing area.

Seldom will a first-aid room be manned on a full-time basis during the second and third shifts; at those times, either a member of the security force or selected line foremen are responsible for administering a certain degree of first aid. If the first-aid provisions are located within the manufacturing facility, several potential security problems are eliminated.

It is also more desirable to have the first-aid activity centrally located.

The Architect and Building Security

The building architect, in most instances, is unaware of the enormous pilferage problem industry is facing. He therefore continues to design structures which in many ways assist the pilferer, frustrate the security director, and deplete the coffers of the industrialist.

A great deal of physical security can be tailored for the building at the architectural planning stage to assist the security director in formulating the physical security plan. There are several areas which should be considered during the design stage of industrial buildings.

1. Openings on the building periphery should be kept to the minimum consistent with operational requirements and emergency evacuation codes.
2. In areas where production processes will be housed, openings on building peripheral walls larger than 96 square inches and less than 18 feet from ground level should be considered for additional protection.
3. Exterior doors intended for emergency exits only should be considered for installation of deterrent alarms monitored at the door location or at the plant security force headquarters.
4. Doors located in remote areas of the building should be supervised by antiintrusion alarms or closed-circuit television cameras monitored at the security force headquarters.
5. Service doors should also be constructed so that the activity in the area can be monitored by closed-circuit television cameras. Service doors on building exteriors should lead directly into the service department so that nonemployee foot traffic is restricted within the protected property.
6. Shipping and receiving doors should be planned so that no gaps exist between the truck trailers and the operational docks. Dock operations should be planned inside the building rather than on the outside where material would be unnecessarily exposed.
7. Shipping and receiving areas should be physically separated by as much distance as possible.
8. Shipping and receiving dock area planning should include provisions for driver waiting rooms equipped with rest-room facilities. Consideration should also include a design to permit drivers to report to the shipping or receiving clerks without requiring that they move through material storage areas.
9. Combustible trash removal from the facility should be planned through a door other than those in the shipping and receiving dock areas. Construction should be such that the trash can be removed from the building without requiring custodial personnel to move outside.

10. Areas within the building that will house classified or proprietary operations should be located out of the way of the normal operational traffic flow.
11. Employment offices and related areas should be located near the periphery of the building to reduce, as much as possible, the area exposed to individuals of unknown character. Doors to these areas should be for the exclusive use of the applicants for employment and not for employees.
12. Employee entrances should be planned so they are located directly off the employee parking lots. These doors should be sufficiently large to accommodate traffic flow intended.
13. Where "clock alleys" would be established near employee entrances, rails or other barriers should be considered so employees can be channeled to the clocks in single lines. This will make identification and control of employees by the security force more positive and effective.
14. Where security forces will control ingress and egress of vehicular traffic from a guardhouse or shelter, the building should be constructed so that the guards will have a 360 degree unobstructed view of the area. Rest-room facilities should be provided, and interior lighting should be controlled by individual manual switches at the location.
15. When office areas are planned as an integral part of the industrial building and these offices will house the administrative operations, openings between office and production areas should be limited to those absolutely necessary for the uninterrupted operational flow.
16. In industrial buildings where it is planned to establish a company store to effect sales to employees, this area should have an outside customer entrance. At the very least, it should be planned to be as close as possible to the employee exit.
17. Cafeterias in industrial buildings should be planned to insure that supplies can be transported to and refuse removed from the area with the least possible traffic through production or warehouse operations.
18. Locker rooms for employees should be planned to insure that they are in close proximity to employee exits rather than located throughout the work areas.

19. Finished-product warehousing areas which are a part of the same building used for production processes should be designed so that these two areas can be physically divided by securing operational doors when warehousing operations have ceased and production operations continue.
20. First-aid rooms should be planned so they are located centrally to the majority of employees.
21. Industrial buildings planned for construction in areas with a high tornado or hurricane incidence should be designed with the least possible glass exposure to the outside. Consideration should also be given to those areas within the building which may be used as shelters by employees in the building during such natural phenomena.

An architect, it can be said, should concern himself as much with providing the physical means to adequately secure his edifice once occupied, as he is in planning basic plumbing.

CHAPTER 7

Security of Shipping and Receiving Docks

It is interesting to note that rising theft losses through shipping operations and the transport of the finished material have increased at such an alarming rate that insurance companies have been forced to increase their premium rates by 20 percent to 30 percent in the last three or four years. Statements from management of insurance companies indicate that premium rates might, in many instances, be reduced by as much as 25 percent if effective procedures were established to reduce losses of the finished product between the assembly line and the local distribution point.

Hopefully, then, the practices and procedures outlined in this chapter, if properly applied, vigorously pursued, and firmly enforced, will contribute to the reduction of losses.

Some of the important factors that should be taken into account are considered graphically in Fig. 7-1.

The degree of security that must be established to adequately protect these critical and extremely vulnerable operations depends upon the type of material being received or shipped. It is readily understandable that if large rolls of steel are received or shipped, the docks would probably not require a high degree of security. However, at a dock receiving or shipping small component parts of radios and television sets, the degree of security must be very intensive and restrictive.

Dock security is directly related to the degree of security which is established at the control points of personnel and vehicular ingress and egress.

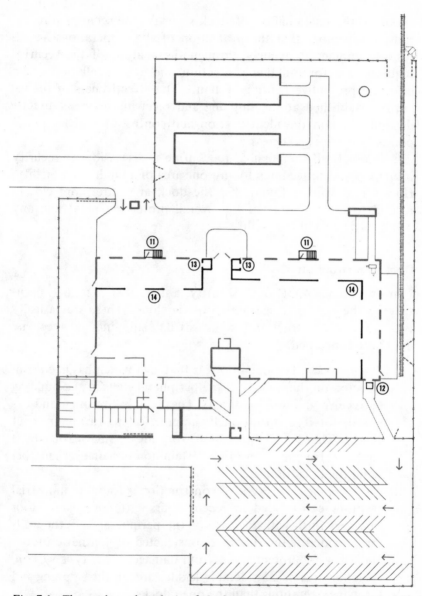

Fig. 7-1. The study and analysis of *dock security* requirements must include the entrances used by drivers (11), the drivers' waiting rooms (13), construction and operational use of overhead doors (14), and most important, the trash removal operations (12).

During the examination of dock security, the security director should remember that the protection of these operations has already commenced, namely through the analysis of the security requirements for building protection and the installation of an effective protective lighting system. The effectiveness of the security established at the shipping and receiving docks is directly dependent upon the identification and control of vehicles at the points of ingress and egress.

It is practically impossible to confine a discussion of security principles and procedures to any one area of an industrial facility. However, it will be found that the dock areas are more closely related to other factors of the overall protection plan than any other single area in the facility.

Dock Construction

In any discussion of dock security, a matter of extreme importance is the type of construction of the dock. There are basically two types of construction insofar as truck shipping and receiving docks are concerned.

1. *Open docks.* The open dock is that area which has been constructed on the outside of the peripheral walls of the building to receive and ship material. These docks are open and exposed, for it is operationally impractical to further restrict them. The dock is protected from inclement weather by a roof constructed to insure a certain amount of overhang beyond the dock floor.

2. *Closed docks.* This type of construction is found in industrial facilities constructed in recent years. It consists of door openings flush with the face of the peripheral wall; the walls are provided with a bumper constructed of synthetic rubber and fiber to protect them from damage. This type of construction creates considerably less hazards in the shipping and receiving operations than the open docks.

The construction of the closed dock is such that virtually no space exists between the truck or trailer bed and the building proper. The vehicle servicing the dock would normally enclose the door opening in much the same manner as if the overhead door were closed.

Trash Removal Operations

Too often the accumulated trash at an industrial facility is removed from the building or buildings through either the shipping or receiving dock, merely because the trash removal truck can be backed up to the dock and the trash easily loaded.

When this occurs, particularly at shipping docks, pilferage will be almost certainly committed. Sometimes it is possible to separate the trash removal operations from the shipping or receiving operations. This is accomplished merely by installing a chain link fence to confine a portion of the dock for trash removal operations.

Even if these security measures are incorporated into the overall protection program, security guards on duty in the dock area must be alert and should make unscheduled inspections of the trash removal operations as they occur. But the installation of a trash compacter at a location remote from the receiving or shipping areas, is the most desirable method, or failing that, baling the trash and removing it by bale over the shipping or receiving docks may be a suitable method.

Driver and Employee Collusion

Herein lies a fertile breeding ground for thievery. Provisions should be made in the dock areas for driver waiting rooms, or areas should be designated to which drivers are restricted after they turn in their paperwork to the shipping or receiving clerk. The drivers should not be allowed to mingle with employees so that they can make arrangements for overloading. If drivers are allowed to mingle with employees in the shipping or receiving areas, collusion in perpetrating thefts will almost surely occur.

If rest-room facilities are not available in the dock area, direct routes to the nearest rest rooms should be clearly marked and drivers instructed not to deviate from these routes.

Any restrictions placed on drivers must be included in the security force's special instructions, and security force guards must be deployed in the area to insure that these restrictions are enforced. The restrictions must also be enforced by shipping and receiving dock supervision.

Housekeeping in the Dock Area

This must receive high priority in the maintenance schedule. All dock areas should be free of anything irrelevant such as trash, old machinery, wooden pallets, and any other material that is not being shipped or received. This will insure that as few places as possible remain for the pilferer to hide items.

The housekeeping at the base of the dock under the trailers should be the same as on the dock floor. Small items are conveniently dropped from the dock into the trash near the wheels of the trailer, and the driver picks up these items under the guise of checking the wheels and tires of his trailer immediately prior to departing.

Storage in Dock Areas

Items are often stored in dock areas merely because more space is available on the dock than is needed for shipping or receiving operations. Items being received should be removed from the dock area into the building, accounted for, and stored as soon as possible. Under no circumstances should any type of material remain in the dock area without being accounted for by the end of the shift during which it is received.

Broken and Damaged Cartons

These must receive immediate special attention where material is being shipped in cartons, particularly when small items are packaged together in larger cartons which are being loaded or unloaded.

When these cartons are broken or damaged immediate action is required either by the security guard on duty or company supervisory personnel in the area to find out what caused it and who is responsible and to fix the blame, if any.

The "accidental" dropping of cartons to break them open is a common means of attempting to pilfer small items from within.

If guards or supervisors do not immediately move to the area where the damaged cartons are, almost certainly some of the items will be stolen by the "careless" employee. It is a common practice for one employee to drop and break open a carton and have several other employees immediately descend upon the broken carton and help themselves to the contents.

A log should be maintained in the shipping or receiving office listing the names of employees who are responsible for breaking open or damaging a carton. The log should contain notations of the day of the week, the time of the day, the identification of the truck being loaded or unloaded, and the identity of those employees who immediately descended upon the broken carton. This will enable an intelligent investigation to be conducted and will determine if any pattern is discernible.

"Hot" Item Rooms or OS&D

The over, short, and damaged areas should be established at docks which are used to ship or receive small, easily pilfered items having a high resale value or of personal use to the employees. This area is also used to store expensive and highly pilferable items that are being shipped or received until they can be either directly loaded or further secured within the warehouse.

All broken and damaged cartons should be removed immediately, inventoried, and stored in this area. Normally the dock supervisor and/or the security guard will be the only person or persons who have possession of the keys to this security room.

An inventory of broken or damaged cartons within the area is extremely important and should be maintained as in a perpetual inventory. The maintenance of an accurate inventory will also insure that none of the items are being removed by personnel on duty at the dock who may be responsible for repackaging the items for shipment.

Drivers Loading Their Own Vehicles

If drivers are allowed to load their own vehicles, it is an open invitation to have property stolen. In many instances, companies will allow the drivers, particularly when they are company employees, to load their own vehicles. Their loads are normally placed on pallets in the dock area and drivers, after checking with the dock clerk, will proceed with the loading.

This practice should never be adopted, or, if in progress, should be eliminated immediately. The time consumed by a security guard or a dock supervisor in checking these drivers to insure that only the designated items are being loaded would almost certainly justify the expense of additional dock employees.

THE USE OF SEALS

A few short years ago, about the only seal available was the metal railroad-type seal. At the present time, there are numerous seals available which are quite adequate in securing the doors on truck trailers. However, the metal railroad seal is most often used.

Whatever type of seal is used it can be effective, so long as it cannot be removed without being destroyed, contains a control number and, preferably, a designation of the organization placing the seal on the trailer. The use of seals on truck trailers, trucks, and railroad cars is an inexpensive method of eliminating theft and pilferage or discovering that these conveyances have been illegally entered.

Too often, however, this procedure is not being followed, seal numbers are not being properly recorded, or the seals themselves are not being properly safeguarded to preclude their surreptitious use.

Sealing Trucks and Trailers

Straight-bed trucks are usually sealed merely by closing and securing the doors with the installed hardware and inserting the seal through the studs and holes provided for this purpose.

Truck trailers, on the other hand, often have rear doors and doors on one or both sides of the trailer. If the side doors of trailers are normally not used, the railroad-type metal seal should be used and the numbers recorded in a permanent ledger and identified with the identification of the trailer. The seal can be left on these doors for many months. However, each time a vehicle sealed in this manner enters or departs the protected area, the numbers of these seals should be examined to insure that the side doors of the trailer have not been opened without authority.

Assuming then, that only the rear doors of a truck trailer will be sealed, this should be accomplished immediately after the truck is loaded or unloaded. If security guards are used in the dock area, they should be charged with the responsibility for security of the seals, affixing the seals, and recording the seal numbers on the shipping documents, as well as in their permanent ledgers.

But regardless of who is responsible for sealing the vehicle, these procedures must be followed, and the individual sealing the vehicle must insure that the seal is affixed in such a way that the locking handles or bars securing the doors cannot be operated without actually breaking the seal. In some instances, it may take more than one seal to properly seal a trailer or railroad car. In that event, more seals should be used and placed together in a "chain" effect.

If a trailer is loaded to accommodate several stops, the sealing procedure can still be followed. First the load is placed on the trailer from front to rear with the material to be unloaded at the last stop loaded first. The individual responsible for sealing the vehicle issues the required number of seals corresponding to the number of stops, and a seal is placed on the trailer prior to its departure.

Assuming that a driver will discharge his load at three different locations, the person responsible for sealing the vehicle would issue four seals to be used for that particular trip, and the seal numbers of the seals to be used at specific points would be entered on the bills of lading or shipping documents pertaining to a specific designation. This procedure is accomplished as follows:

1. When the truck is completely loaded, a seal is placed on the door and the number recorded on the shipping documents which will be delivered with the first portion of the load. This number is the seal number checked by the security guard at the point of egress.

2. When the vehicle arrives at the first point of destination, the person responsible for receiving his portion of the load checks the seal on the truck against his documents to verify that the seal is intact.
3. When the portion of the load has been discharged, this person affixes the next seal which is indicated on the shipping documents pertaining to the second stop.
4. This procedure is repeated through the second and third unloading points, and at the third unloading point, the truck, now empty, has the final seal attached. This is the seal that should be on the vehicle when it returns to the point of origin and is checked by the security guard at the control point.

A number of variations to this procedure can be adopted to insure that the vehicle involved does not move any distance without a seal being attached. Sealing assists immensely in determining whether or not a driver of a vehicle is making stops and unloading material which has been loaded on the truck without authority.

Empty truck trailers leaving the dock area should be sealed immediately after the truck has been unloaded. The number of this seal should also be recorded, and the security guard at the point of ingress should then check the seal number to insure that the vehicle has not been loaded with unauthorized material between the dock area and the point of egress. This procedure not only eliminates the possibility of loading items on the vehicle but speeds up inspections at the gate house.

Sealing Railroad Cars

All railroad cars arriving inside the protected area will normally contain the seals of the railroad company. These seals should be checked by the security guard at the point of control, and a notation made of the number and whether or not a seal was actually attached.

If railroad cars are to remain on a siding inside the protected area without any further protection, it may be advisable, depending upon its contents, for the security force to place a padlock or similar locking device on the car as additional protection until it is spotted for unloading.

Railroad cars should be inspected and the doors closed and sealed as soon as they are unloaded.

If members of the security force will be responsible for sealing railroad cars, provisions must be made, in most instances, to have dock employees close the doors. The majority of boxcar doors cannot be closed easily by one man.

Extreme caution should be exercised in this procedure to insure that the inspection of the car is made immediately prior to the doors being closed and the seal affixed.

There have been instances where a security guard has inspected the entire interior of a railroad boxcar and found it empty. The doors were closed by dock employees at a later time while the security guard was conducting other duties. When the guard returned, he found the doors closed and placed a seal on the car without knowing whether or not material had been placed in the car during his absence and the door closed.

Security of Seals

A procedure which may be used to insure that all seals are being properly handled is to issue the seals in blocks of numbers in numerical order. The seals are logged in on a log book when received and, to insure every seal is accounted for, a notation of the identity of the vehicle on which the seal was placed, or to which vehicle assigned, should be noted in the log book.

Unused seals must be secured in a locked container after docking operations have ceased for the day. Seals should be accounted for either by the security force or the dock supervisor on a daily basis or by shift, if operations continue on a twenty-four-hour-a-day basis. Broken or damaged seals which will not be used and those seals removed from the transport vehicles at the docks should be recovered and immediately placed in a secure container until they can either be destroyed or removed from the property.

Loaded or Partially Loaded Transport

Transport vehicles which remain on railroad sidings or parked inside the protected area should be further protected by using padlocks to secure the doors, if they are to remain in the area for any length of time.

Padlocks would normally be placed on the conveyance by the security force and may be removed either by the security force when the truck or railroad car departs the protected area or by the supervisor who is responsible for completely loading the partially loaded transport the following day.

Trucks remaining within the protected area for any length of time must be parked a sufficient distance from the chain link perimeter barrier to insure that persons on the outside of the barrier cannot approach the back of the truck, cut a hole in the chain link fence, and attack the trailer without being seen. Security patrols through truck parks must be designed to insure that each patrol includes an inspection of all transports.

SECURING THE TRUCK PARK

When trucks will routinely be parked and remain for some length of time in the protected area after they have been loaded, specific procedures for this protection must be developed. Some measures that may be adopted are:

1. Parking areas for this type of vehicle should be well lighted and isolated from any routine employee traffic.
2. Loaded trucks should remain within the primary protected area and not parked in employee parking lots.
3. Parking areas should not be located near vehicular gates merely for the convenience of drivers.
4. Trucks which are loaded with material of high value or easily disposed of if stolen, such as cigarettes, shoes, clothing and household appliances, should be parked, whenever possible, well within the protected area and, preferably, in a position where it can be under constant observation of the guard at a fixed post.

5. To further deter thefts of entire trailer loads, locking devices are available that can be placed over the trailer kingpin and locked in place to deter hit-and-run tactics from being used.

LOAD SECURITY IN TRANSIT

Company truck drivers who are involved in long, over-the-road hauls should be instructed to make overnight or rest stops only in well-lighted areas. They should also be cautioned, even if they sleep in the cab of the vehicle, to keep cab doors locked.

Additional precautions should include furnishing the driver with padlocks if he habitually parks overnight or makes long rest stops using available facilities along the way.

An excellent procedure to insure that maximum security to loaded trailers is not overlooked by the driver is for the company to establish a policy requiring that all doors on the trailers remain locked, in addition to being sealed. The driver would be issued a key to be secreted in the tractor of the vehicle and not placed with the ignition or door keys.

Safety during loading and unloading operations at truck and railroad docks should be included in all operational procedures established. If security guards are being used in these areas, their responsibilities should include inspection of the vehicles to insure that trailers and railroad cars are properly chocked.

Experience has proved that accidents frequently occur when forklift vehicles are used to load and unload truck trailers. Air brake failures do occur, due to air leaks, and the weight of the forklift truck moves the trailer from the dock. Needless to say anyone involved in this type of accident is almost always killed or permanently disabled.

CHAPTER 8

Locks, Key Control, and Security Containers

After examining the existing protective lighting system and completing a partial analysis of additional lighting requirements without having to acquire the services of an illuminating engineer consultant, the reader has surely developed a certain amount of self-confidence. Similarly, to obtain a fully effective locking system through the proper selection of locking devices does not require that the reader be an accomplished locksmith. At times, however, it may be necessary to use the services of a locksmith to insure that locking devices are properly selected and professionally installed.

In the past few years, vast improvements have been made in the manufacture of every type of locking device. More substantial, tool-resistant metal has been developed, new innovations have been incorporated into the lock plug and, in one particular instance, the basic concept of the pin tumbler cylinder has been reengineered. This particular locking device utilizes three rows of key pins containing a total of at least twelve pins as opposed to the conventional cylinder. The conventional cylinder has its key pins equally spaced and in one single row only, and usually contains only five or six pins. See Fig. 8-1.

The key-operated lock is the most widely used—and the simplest to defeat. There are numerous ways to defeat these key locks, all of them easy to an expert. George Hunter, a journalist in the fields of crime and self-protection, has said, "I'm quite unskilled with a pick myself, but I can unlock most pin tumblers in fifteen minutes

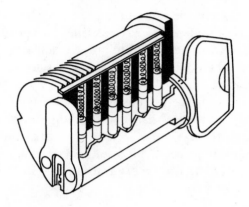

Courtesy of Sargent & Co.

Fig. 8-1. The six-pin tumbler lock offers a degree of "pick resistance" that
is acceptable to some security requirements.

without a key. Some very skilled people can do it in a second
or two."[1]

One should recognize that locks cannot be judged by their size,
weight, or appearance. Although a lock appears to be of substantial strength, it may offer little or no security. The working
parts inside the lock contribute a great deal more to the effectiveness of the device than merely its size or weight.

To understand the values of different types of locks and locking
devices, one should be somewhat familiar with the relative security
of and the features of locks commonly used. One should know
something of the protective features that the lock offers against
the time it takes to pick it. As an example, the amount of security
offered by a disc tumbler is about three minutes, since it takes this
time, or less, for an expert to pick this type of lock. Therefore,
if a three-minute delay does not prove to be the degree of security
which is desired, the disc tumbler lock is not adequate for the job.

The security of key type locks ranges from none to excellent. A
lock has no security value when it can be opened without a proper

---

[1] George Hunter. *How to Defend Yourself, Your Family and Your Home.* New York:
McKay, 1967.

key in approximately the same time it would take to open it with one. A lock is considered to have excellent security value if the lock must be forced or destroyed to quickly defeat it.

This type of lock for a particular area is selected in much the same manner as an alarm device would be. First, the degree of security required must be established and then the proper locking device selected. Note also that it would be pure folly to install a locking device with maximum security potential on a door constructed of two thin pieces of plywood tacked to a wooden frame. The entire door offers little more resistance than the strength of these thin plywood panels!

Types of Locks

Locks of any type are considered merely as delaying devices. Before selecting a lock, it is necessary for the security director to familiarize himself with the types of locks commonly in use today. He should also be familiar with the various types of locking devices and their general capabilities and limitations. Below is a brief discussion of the types of locks available and their general security value.

Warded locks have no security value. They can usually be identified by a single plate covering which includes the doorknob and the keyway. This type of lock was commonly used on doors in residential buildings constructed prior to World War II and is still in use on older residences in rural areas.

Disc tumbler locks are used in most automobile doors and are often found on filing cabinets and desks. They are generally made of soft metal and can be identified by the keyway, which does not rotate in the locked position. The flat or concave edges of the discs can be seen in the keyway. The concave effect on the disc is often caused by wear through use because of the soft metal. This lock is very similar to the warded lock, and it provides more privacy than security.

The conventional *pin tumbler lock*, consisting of five or six key pins, will effectively delay intrusion from approximately two to five minutes, depending upon the number of key pins involved, the tolerance permitted in manufacture, and the expertise of the individual picking the lock. The lock is extensively in use throughout industry and can be identified by the keyway, which is irregular and eccentric in shape and requires a key with grooves along both sides of the shaft to operate it.

Circular keyway locks have a circular keyway, containing a slot that guides the key into the keyway properly. The key pins are located at the bottom or far end of the keyway and are depressed to the correct depth when the proper key is used. When the key pins are uniformly depressed, they are aligned with the shear line of the device and the lock is then operated. This lock is extremely difficult to pick and has many applications in installing a secure locking device system. This type of cylinder applied to a padlock is illustrated in Fig. 8-2.

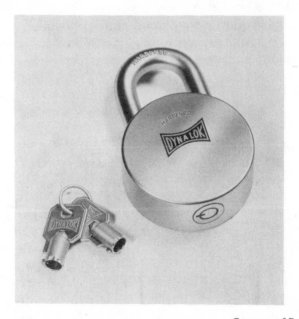

Courtesy of Dynation Corp.

Fig. 8-2. The tubular lock cylinder installed in a padlock offers excellent security at perimeter barrier gates.

There is a wide range of *lever locks* manufactured and therefore they offer different delay times. High-grade lever locks are used on safe deposit boxes. Simple, easily defeated lever locks are found on chests and cabinets. The security of this type of lock is based on the number and type of levers built into it. The lock can usually be identified because of the unobstructed keyway, which will rotate approximately 25 to 40 degrees in either direction without opening the lock.

Maximum Security is the trade name of a locking device introduced by Sargent & Company, New Haven, Connecticut, in 1965. This locking device is part of a system known as the Sargent Maximum Security System. Maximum security is obtained from the three rows of pins which interlock in the keyway, giving these cylinders unprecedented pick resistance. The keys operating these lock-devices are six-sided and reversible. (See Fig. 8-3.) Each system has its own computer-scrambled combination of pin lengths and positions. The pins in the lock converge on the key from three different angles. The notches and grooves common to the conventional key have been replaced by a number of precisely milled hollows on both the edges and flat surfaces of the key.

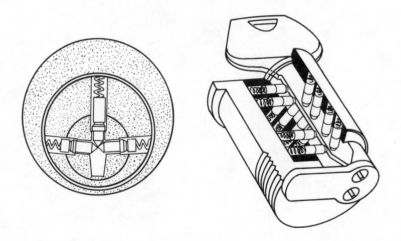

Courtesy of Sargent & Co.

Fig. 8-3. The Sargent Maximum Security Lock with its three rows of pins offers a formidable barrier to the "lock picks."

This type of lock is excellent for use on perimeter gates and peripheral doors of the building. One should keep in mind that the lock should be installed on doors of substantial construction.

Combination type locks are difficult to defeat successfully, and the delay time will vary with the expertise of the intruder. To obtain any degree of security from a combination type lock requires that at least three combinations be properly aligned before the lock is opened. The lock should also have the capability of changing the combinations in a short period of time. When the use of combination locks is required in providing security for government-classified information and material, the regulations provide that the combination lock must contain at least three combinations and must have the capability of combination change.

Padlocks are manufactured with any one of the locking methods discussed above. The delay time in defeating a padlock depends entirely upon the following conditions:

1. The type of padlock installed
2. The hardness of the metal shackle and the construction of the body of the lock
3. The material with which the hasp is made and the manner in which the hasp is installed
4. Whether or not both ends of the shackle are secured when the shackle is in its closed position. (See Fig. 8-4.) If just one end of the shackle is secured, the lock is easily defeated by "tapping" it open.

Cypher locks provide a considerable delay, because no keys are required to operate the device. The locks are manufactured with a series of buttons, either lettered or numbered, and can be opened only when the buttons are depressed in the proper sequence. The cypher lock is manufactured to be operated either manually or electrically. The combination of these locks can be changed quickly. A good quality cypher lock offers a high degree of security, since no keys, which could be compromised, are issued, and combinations issued to individuals no longer having a need for them can be eliminated merely by changing the combination.

Courtesy of Sargent & Greenleaf, Inc.

Fig. 8-4. The "protected shackle" and two lines of "pins" afford maximum security in a padlock providing the hasp is properly installed.

When considering the cypher lock, the security director must remember that any locking device in which numbers are memorized can also be compromised, much the same as keys can be compromised. However, the calculated risk of such compromise is considerably diminished. An illustration of a manually operated cypher lock appears in Fig. 8-5.

There are literally hundreds of *miscellaneous locking devices* on the market, and they may be used in various ways. The security value of most of these devices ranges from none to fair. Some of the devices in this category which are in common use are telephone locks designed for both the dial and touch tone telephone (see Fig. 8-6), locking door chains which can be key-operated, window locks, and a multitude of other devices. These devices have little security value and are manufactured primarily for residential security needs.

Courtesy Simplex Security Systems

Fig. 8-5. The cypher lock does not require keys to be issued. Combinations can be changed by the guard force in minutes, should any question of compromise arise.

Selecting a Locking Device

The preceding discussion on the types of locking cylinders manufactured and the degree of security which can be expected from them must be kept in mind when selecting the actual hardware that will be installed.

The selection of the proper locking device to be installed, especially on openings along the periphery of the building, will often determine whether or not an intruder can be successfully deterred. If entry into the building is made as difficult as possible, an intruder will often bypass a choice target for one of less interest to him simply because a successful attack is easier. Following is a discussion of locking devices available which, if fitted with the proper type cylinder, will complement the overall protection plan of the facility.

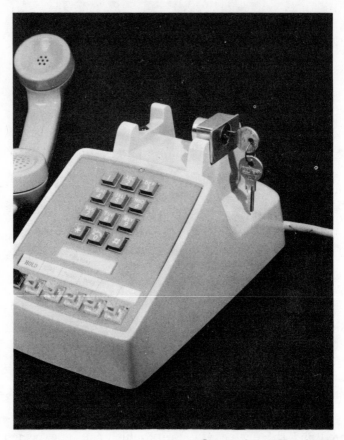

Courtesy Loxem Manufacturing Corp.

Fig. 8-6. Even the touch-tone telephone can be secured against unauthorized
use.

Double cylinder locking devices are devices which are installed
on doors. They permit the lock to be opened from either side of
the door. This device requires that a key be used to open it from
either side. It is particularly effective in securing doors containing
glass panels because it does not allow breaking the glass, reaching
inside, and operating the lock by merely turning a thumb lever.
These devices are commonly referred to as "security locks" and
recommendations for their use were discussed in Chapter 6.

Jimmy-proof locking devices have a vertical throw bolt as opposed to the normal horizontal throw bolts of most other locks.

This device is considered jimmy-proof because a favorite method used by burglars to force entry is spreading the doorjambs by using an automobile jack and two pieces of wood or other material. If a short horizontal throw bolt is used, the doorjamb can be separated from the throw bolt and the door opened. The vertical bolt locking device secures the door to the jamb, preventing the possibility of spreading. "Spreading" means merely increasing the distance between the door and the doorjamb, thereby, defeating the lock.

Emergency-exit alarm locking devices are necessary because emergency exits cannot be secured with locking devices that can be opened only with a key; this would delay emergency egress. Locking devices can be installed on these exits with panic-bar hardware which, when depressed, releases the locking mechanism. These devices are equipped with a self-contained dry-cell battery which activates an alarm sounded locally when the panic-bar release is depressed and the lock is opened. This device can be opened without activating the alarm by using the proper key to bypass the alarm activator.

Recording locking devices contain mechanisms which provide a tape-printed record each time the lock is opened or closed with the proper key. In addition to recording the time the lock has been opened or closed, it will record the particular key that was used and provide after-the-fact information on the printed tape which can be used for various purposes.

Sequence locking systems may be effectively used throughout office areas in the facility and will insure that every door on which the sequence lock is installed is secured. It is physically impossible to "forget" any door in the system. The completely mechanical system requires the locking of each door in a predetermined sequence and prevents re-entry except through the main entrance door. The main entrance door may or may not require additional protection. See Fig. 8-7.

Electromagnetic locks are installed on the side of the door facing the protected area. The device is electrically operated and, generally speaking, consists of the electromagnetic lock, the strike plate, and the power control unit. (See Fig. 8-8.) The lock is activated electrically; when activated, it is said to have a strength of one-half ton holding force.

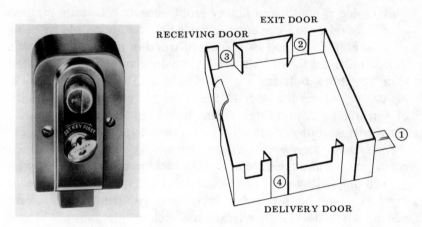

Courtesy of Simplex Security Systems

Fig. 8-7. The sequence locking system insures that all locks in the system are engaged before the last door is secured. In the sketch at right, all doors must be secured before the basement door (1) can be locked.

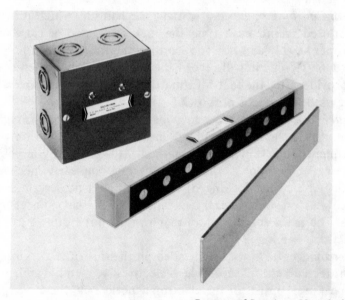

Courtesy of Securitron Magnalock Corp.

Fig. 8-8. The "Magnalock" works on the electromagnetic principle, has a 1,000 pound pull guarantee and can be operated from a remote location by engaging or disengaging the electrical power.

These locking devices are effectively used when a fixed security guard post is some distance from the door opening, and the security guard has a means of identifying the individual desiring entrance. The power control can either be button or key-operated.

Card-operated locking devices are electrically energized and usually operated on the electromagnetic principle. A card, rather than a key, operates the lock. This card can be used for dual purposes, because it can be manufactured as a personal identification card and also operate a lock or locks to a particular area.

Card locks are used where the degree of security required is such that a key-operated lock would not furnish the necessary delay time. The operating cards are impregnated magnetically and coded to a particular lock. When the card is inserted, it is electronically read and the lock is electrically operated.

Electric strikes are installed in the doorjamb and are electrically operated from a remote location. The door being protected should always be equipped with an automatic door closer; the door remains locked at all times that it is closed. Individuals desiring entrance will announce their presence through a buzzer system or speak with the door operator through an intercom system from the outside of the door to the controlled area. Before the person is authorized to enter, he must be properly identified, and then the strike is electrically released to permit him to open the door. See Figs. 6-1 and 6-2.

Removable core locking devices are manufactured so the core of the lock can easily be removed and replaced without the necessity of removing the entire locking device to accomplish rekeying. This lock can be identified because the cylinder is shaped like a figure 8. The system provides for operating keys and core keys. A "core key" is used to disengage the core or cylinder from the locking device and replace it with another core immediately, with the same result as rekeying the fixed type cylinder.

Examination of individual locking devices using *horizontal throw bolts* should be detailed enough to insure that the throw bolt being used is long enough to furnish the protection required. Often horizontal throw bolts are short and, because of the distance between the door and the doorjamb, as little as an eighth of an inch of the throw bolt may actually secure the door to the jamb.

Obviously, this condition is easily defeated by spreading the door from the jamb using practically any type of leverage device, i.e., a jimmy, as described earlier.

Key Control

If an effective locking system will be or has been installed, control of the keys and an effective key record must be maintained, or the entire locking system will surely be compromised. Proper key control is simple to establish and maintain if incorporated into the security program at the time the locking devices are installed. Once key control has been lost, it is difficult and time-consuming to reestablish. Without absolute control of every key to every locking device, the system soon deteriorates and the degree of security originally established is questionable or lost.

An effective key control system consists of two primary elements:

1. A key cabinet of sufficient size and well constructed, properly installed and secured, to hold the original key for each lock in the system and such additional keys as may be required. See Fig. 8-9.
2. A card file or other administrative means to record key code numbers and indicate to whom keys to specific locks have been issued.

The simplest method to *reestablish key* control when the security of the locking devices installed throughout the facility is questionable and no key control system is available, is to purchase a key control system including materials for administrative records and other items needed to assist in assembling and identifying the keys available.

There are many excellent key control systems available on the market today. All include descriptive literature outlining the procedure necessary to reestablish key control.

Economical Changes

If major changes by replacing or adding locking devices to the system are necessary, some economy may be realized or a reduction of the initial investment accomplished if the locking devices are

Courtesy of P. O. Moore and Co.

Fig. 8-9. Key control is essential in a secure locking system. All keys must be controlled at all times. Administrative records in the key control system must be kept updated. A secure key cabinet is mandatory.

changed over a given period of time. The cost of changing all locks at one time may be prohibitive.

Another factor to consider is that if the locks on the exterior gates and doors of the facility are the first ones to be changed, those locks removed may well be used as locking devices on the interior of the buildings in areas requiring a lesser degree of security.

The locking devices on the exterior doors and gates are of paramount importance in establishing a secure locking system, and consideration must be given to changing all of these locks first. Then, if a study is conducted of each individual area requiring a locking device, and a schedule is established for the provision of new locks for the system, the cost of purchase of new locks can be kept to a minimum, and yet the integrity of the entire locking system is increased.

If key control is established and key inspections conducted at least semiannually, and keys are issued only to responsible individuals who actually have a need for these keys, an adequate, effective locking system will have been established.

Master and Submaster Systems

A study of all locking devices used throughout the facility must be conducted in relation to the operational plans in order to arrive at a workable master and submaster system. The number of such systems will depend upon the size and complexity of operations and the facility.

To facilitate the study and analysis of the submaster systems required, it will be helpful to understand various key designations:

1. The *change key*, sometimes known as the "door key," is an individual key to any single lock in a master keyed system. Any lock having a given number of tumblers can be "changed" several times. Therefore, the term "change key" is used to designate the individual key belonging to a particular change of the lock.

2. The *submaster key* was referred to some time ago as the "floor master key," because it was issued to the housekeeping staff at public hotels. This key will open all locks within a particular subgrouping. As an example, a submaster group at an industrial facility might be all the locking devices used to secure only the warehousing areas, the administrative office areas, or other areas where all of the activity within the area would usually occur during the same period of time.

3. The *master key* is established where two or more submaster key systems are involved. The master key, obviously, would open any of the submaster systems within the group.

Normally, a system in an industrial facility terminates with the master key or the master locking system. If it is necessary to expand the locking system to a greater degree, a number of master systems would be established. When such systems are established, a grand master key, a great-grand master key, etc., are then established.

In any locking system where master, grand master, or core keys are used, additional security for these keys must be considered, because the compromise, loss, or theft of any one of these keys would immediately reduce the security value of the system and probably be extremely expensive to replace. Wherever possible, grand master, master, and core keys should be issued grudgingly. These keys should be secured in a safe containing a combination lock.

The higher classification the locking system has, the thinner the master and grand master keys will be. It is not infrequent that a security guard will break a master or grand master key, primarily because it has been machined to the point that it is very thin and weak, and then it is additionally worn through constant use!

Security Containers

There are basically three types of security containers: vaults, safes, and filing cabinets.

The *vaults* installed at an industrial facility are normally used for the storage of records, computer data tapes, and material of this type rather than for the storage of high value items, monies, or other negotiable material or documents.

Safes can be purchased in numerous sizes and weights. There are also numerous types of safes available, and the type chosen will depend upon the particular need. A special type is shown in Fig. 8-10. Small safes mounted on casters or rollers could be easily removed from the premises; these are often chained to an eye embedded in a cement floor, or another such arrangement is made.

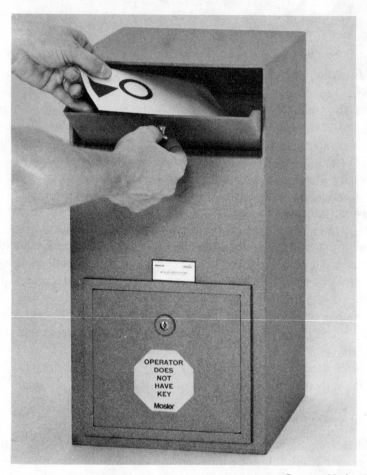

Fig. 8-10. Monies accumulated by delivery truck operations can be better protected if a "dropository" is permanently installed in the delivery vehicle.

It is a common practice today, particularly where monies or material of extremely high value are being stored, to install the safe in the floor of the office. Safes are manufactured for this specific type of installation; they offer a great deal of resistance to attack. The entire body of the safe is embedded in a great deal of cement and only the door with its combination is exposed.

Filing cabinets are the third type of security container, and this is the type most often found in industrial complexes. Filing cabinets of various sizes and designs are available. Normally, if government classified material is not being stored, a filing cabinet with a rating of one-hour exposure to fire would suffice. If this type of cabinet is fitted with a three-way combination lock and further protected by an alarm device installed specifically to protect the container, or a space alarm that protects the entire area, this container furnishes excellent protection.

If classified government work is in progress and highly classified documents are being stored, in most instances the type of filing cabinet described will not suffice. Again, reference must be made to the Department of Defense manual, *Industrial Security Manual for Safeguarding Government Classified Information*, also available from the U.S. Government Printing Office. Little would be gained by discussing the security requirements outlined by these regulations, since they are subject to change and, if an industrial facility is conducting government classified operations, the responsible official will detail the specific security requirements involved in that particular classified contract.

When determining the requirement for security containers at a facility, the degree of security required must, obviously, be established first. In many instances, the use of a filing cabinet similar to that described above, and the selection of an adequate security alarm will provide adequate protection. The decision on which type of alarm to employ can be based upon the information contained in Chapter 11.

CHAPTER 9

Identification and Control of Personnel

Identification of personnel starts when the job seeker enters the employment office and is completed only after the prospective employee has been thoroughly investigated.

Background investigations of all employees must be made if pilferage, thefts, accidents, and incidents are to be avoided. The depth to which the investigation goes will depend upon how sensitive is the position being filled, a research and development employee should be more thoroughly investigated than a production line hourly applicant, for example.

As a minimum, the following items should be obtained or investigated:

1. Police record, where possible, in all places of residence for the past ten years.
2. Past ten years' employment, or unemployment, to cover all periods in excess of thirty days.
3. Three personal references who have known the applicant for at least five years should be secured and interviewed.
4. If military service falls within the ten years being investigated, the last DD Form 214 (Military Qualification Record) should be examined.
5. A medical examination should be conducted by a doctor selected by the company.

When the investigation has been completed and the applicant found satisfactory as regards character and job qualifications, he then becomes a member of the group of employees who will be discussed in the following paragraphs.

Controlling the movement of personnel from the time they enter the plant until they depart is one of the most important facets of the security program. The establishment of a perimeter barrier and gates breaching the perimeter barrier will assist in control of personnel. (See Fig. 9-1.) However, other aids and techniques must be incorporated into the security plan in order to properly identify those personnel who are authorized to enter and exit any specific point at a given time.

Regular employees and other authorized persons should be able to enter the premises without undue delay, but at the same time unauthorized persons must be excluded or permitted to enter only after being properly identified and the fact has been established that they have business within the protected area.

It is obvious, then, that identifying and controlling personnel at the points of entry is to insure that only authorized persons are permitted in the protected area. However, the methods initiated must not impede the operational efficiency at the facility. The identification system used will depend upon many factors: the size of the facility; the number of personnel entering through a particular point during a given period of time; the number of security guards available to control these personnel; the various categories of personnel; and the overall degree of security established for the facility.

The system established must be workable and as simple as possible, but still effective enough to insure that the proper degree of control is established to eliminate the entry of unauthorized persons. If the system is effective, it will greatly reduce unlawful acts that may interrupt the operations.

Personal Recognition

Personal recognition is the most positive means of identifying any individual. In small industrial plants and businesses where a relatively small number of persons are employed, it is possible that a gate guard could personally recognize each employee who

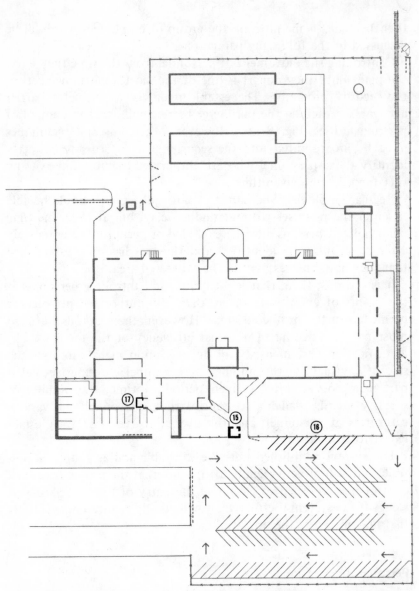

Courtesy of Integrated Communication Associates

Fig. 9-1. Identification and control of personnel includes establishing control points at the main gate house (15), visual control of visitors parking (16), and at the switchboard operator/receptionist location (17).

works at the plant, if he is regularly assigned and remains on a particular shift.

Past experience has proved that an individual guard cannot personally recognize over 30 or 40 persons per shift with any degree of certainty. Where the turnover rate of employees is high, personal recognition to identify employees is almost ineffective. Considering employee turnover, increases and decreases in the number of employees from time to time, and to insure that visitors and nonemployees do not enter the protected area, some type of identification system must be established.

Where no employee identification system exists, or if the employee identification system has been allowed to deteriorate and is being revitalized, an educational program for employees should be established prior to instituting a personal identification system. Past experience has proved that employees resist an identification system for no apparent reason other than that a certain percentage will assume the attitude that they are no longer being trusted. This, obviously, is not the reason for establishing an identification system; but rather the opposite, for the establishment of such a system insures the employees that anyone not authorized is not allowed to enter the facility.

If a proper educational program is launched, the reason for the system being established made clear to the employees, and this information disseminated to all employees, the establishment of a system of employee identification will be accepted with little or no objections from the group as a whole.

Types of Identification Media

There are basically two categories of identification media which are acceptable in establishing personal identification of individuals authorized to regularly enter and leave the area.

The *identification button* usually is a round or oval button containing the name of the facility and a control number. The personnel office issues this when an employee is hired. The employee may retain the button as long as he is employed, or he may be required, at smaller operations, to turn it in at the guard post each time he departs the property.

The button type media offers very little security whether or not the button is retained, since, unless the bearer of the button can be personally recognized or identified in some other manner, there is no way to identify him as being the person authorized to carry the button. An employee who is fired could pass this button to anyone else or use it himself to gain unauthorized entry. Again, if the button is lost and the guard does not otherwise identify the bearer, an unauthorized person may be admitted.

Button identification can be effectively used to control personnel on a temporary basis. As an example, if construction work is in progress or a contractor is working within a protected area, a group of numbered buttons of a particular color are issued to the contractor who, in turn, issues the buttons to his employees or the employees of subcontractors. He then furnishes the security guard with a roster of the names and numbers of the buttons he has issued. This affords the guard an opportunity to make spot checks on these particular employees. However, the button identification media offers little security when used on a permanent basis.

Laminated card identification is the best means of establishing identification of personnel. The card must be laminated in such a way that it is tamper-resistant. The paper used should either have an intricate background design or ink or dye which would be affected if heat or solvents were used in an attempt to tamper with the card. The paper should be noticeably affected if an eraser or ink eradicator is used to change the data on the card. Equipment for producing cards is shown in Fig. 9-2.

The laminated identification card should contain the following, if it is to be considered effective over an extended period of time:

1. The name of the company represented
2. A recent color photo of the bearer
3. The printed name and written signature of the bearer
4. The date of birth, height, and weight of the bearer
5. An identification control number or social security number
6. The title, and signature of the person authorized to validate the card
7. The date of issuance and, if appropriate, the date the card will become void.

Courtesy of Instant Identification Systems Corp.

Fig. 9-2. Equipment for producing identification cards consists of five basic items: (1) camera, (2) tripod, (3) and (4), card and photograph die cutters, and (5), laminator. Many other accessories such as the economy models of a card cutter (left foreground), and corner rounder (right foreground) are available but are not necessary to produce a laminated identification card.

In certain circumstances, cards, such as the military identification card, bear fingerprints or thumbprints to further establish identification. Normally, however, the prints do not serve any useful purpose in industry and can be eliminated.

Display of the identification is of utmost importance. Employees should be required to wear the badge in a designated place on their outer clothing—or they will be wearing them on their belts, hooked to their trouser pockets, or under their outer clothing. If the badge is not worn where it can be readily seen, ingress and egress will be slowed up through the guard having to ask to see badges.

Color-Coding

The identification button or laminated card, as the case may be, is color-coded by using different colored buttons or cards, or perhaps a different color background in the picture used on the laminated card.

Color-coding is used where higher security requirements for the control of personnel within the protected area is necessary. As an example, employees who are authorized to work in the warehouse may be issued a green identification card, the production workers a red one, and research and development personnel a yellow one. With color-coding, the security guards patrolling the inside of the plant can quickly differentiate between authorized and unauthorized personnel, if the card is being worn as required. If an effective educational program in personnel control is established, employees, particularly supervisors, can help in seeing to it that employees keep to the rules as regards security restrictions.

Color-coding can be extended to almost any degree desired. Take for example above where the cards are red, green, and yellow. The plant manager may be authorized to enter all three restricted areas. Instead of carrying three cards, he would merely carry one card, probably white with three small blocks on the front of the card, one green, one red, and one yellow, which would authorize him to be in any of the three restricted areas.

In plants with a complete security program, the color code could even be extended to color-coding the emergency evacuation doors used by particular departments and color-coding the aisles leading from the department to these doors by painting the aisles the color assigned to that department.

To extend the color-coding even further, outside areas where the employees who have been evacuated will be assembled could also be color-coded merely by placing colored signs or stakes in the area where they are to assemble.

If the security requirements are such that color-coding is used inside the plant, management would not want the employees to mingle outside the plants once they have been evacuated during emergency evacuation drills and actual emergencies. This would give them the opportunity of passing items on, because they would have departed their areas without being checked by any security guard.

Identification Systems

There are three basic types of identification systems. They can be used for long periods of time if properly safeguarded and will effectively control all employees' ingress and movements within the facility. One system or a combination of any of the systems may be used. The one selected will depend upon the degree of security that is required.

1. The *single system* is usually employed in the average industrial plant. It consists of issuing a laminated card to each employee authorized inside the protected area. This employee usually carries the card with him at all times. The card may or may not be color-coded, depending upon the security requirements.

2. The *exchange system* requires the use of two cards for each bearer. Each card contains identical data, including pictures and vital statistics, but the remainder of one badge is usually different from the other in color and/or design. One of the cards remains in the possession of the bearer and is handled in the same manner as under the single system. The other card remains at the entrance to a specific area where it is to be used.

When the bearer arrives at the specific area, he presents his issue card to the guard on duty at the entrance. The guard issues him the second card after the bearer has been identified and has surrendered the first card. In order to enter the restricted area, even though the card authorizing entry has been issued, the bearer must sign in on a register indicating the time he entered. When he leaves, he surrenders the identification card for that particular area, picks up his original card, signs out on the register, and departs.

3. The *multiple system* is merely a further development of the exchange system. Under this system, there are two or more card exchange areas as described above, with a special card filed at each special area.

Control of Visitors

The above discussion prescribes the means of identifying and controlling employees and others who regularly enter or leave the protected area. Obviously, any type of industrial installation will have visitors who will be authorized to enter from time to time for certain reasons and specific periods of time. Individuals authorized to enter on a temporary basis must also be identified and controlled, or the security plan will be compromised.

Control of visitors in many industrial facilities, is the responsibility of the receptionist, telephone operator, or a person who performs both these duties when visitors are channeled through the main plant office. The following discussion applies equally to the receptionist and to the security guard on duty at the gate if that is where visitor control is established.

There are numerous different methods of controlling visitors to a facility and, again, the method used will depend upon the degree of security required, the size of the facility, the frequency of visits and the number of authorized points of ingress. Therefore, this discussion will be confined to basic visitor control. The restrictions may be upgraded or downgraded as the security situation dictates.

Reserved parking spaces for visitors should be established as near as possible to the point of entry. These parking spaces should be within the area of observation of the security guard where possible. Sufficient signs should be erected to insure that the visitor does not wander around the property looking for the guardhouse or the authorized point of entry. The route from the parking area to the point of entry should be the most direct route.

When the visitor approaches the control point, presents his credentials, and is properly identified, the person whom he wishes to see should be contacted to determine whether or not he will see him. The visitor then should be documented.

A visitor's log for all nonemployee personnel authorized to enter the protected area must be established at the authorized points of ingress. It should contain at least the following information:

1. Full name of visitor and company represented
2. Reasons for the visit
3. Person in the plant being visited
4. Time of entry
5. Time of departure
6. A column for remarks, including a list of any safety equipment, such as safety glasses or hard hats, issued.

To escort or not escort the individual will depend upon the degree of security being planned. If an escort is required, he should meet and escort the visitor throughout his stay in the protected area. In some instances, depending upon local ground rules being established, or perhaps on which restricted area the visitor will be authorized to enter, it may not be necessary for the visitor to be escorted.

A visitor's pass or badge should also be issued, so the individual can be identified while he is inside the protected area. This identification may be in the form of a visitor's button with a control number on it, or it may be a laminated card with the name of the company and the word "Visitor," or any other type of identification media which would identify him as a visitor to the facility.

Where the security requirements will allow, an excellent method of controlling visitors is to use a "sales slip type register." This is a register similar to that used in hardware stores and in old general stores to complete the sales slip. Slips or passes are preprinted with the amount of information necessary at a particular facility. The form is in duplicate in the machine. The guard or receptionist completes the necessary information required by the form, has the visitor sign it, operates the handle on the machine, and detaches the original copy.

The duplicate copy remains in the machine as a permanent record. The original copy is usually folded and placed in a plastic holder, and the visitor is required to either pin or clip this plastic holder to the outer clothing. The departing visitor is required to depart through the same point of control, and the pass is recovered. Visitors' passes including the sales slip type of pass may be color-coded.

With the sales slip pass, a visitor's register is not necessary. The method is a quicker, more positive means of control than issuing a button or card and registering the visitor on a log.

Group Tours

Group tours are often conducted at industrial facilities as part of a community public relations program. Tour members are not normally registered individually. However, the person in charge of the group should definitely be registered, and the total number of the group should be indicated on the pass or on the register. The security guard or the escort for the group should count every individual in the group, and verify the number with the count in the register. When the group departs, the identification issued to the leader of the group should be recovered, and all members of the group counted to insure all have departed the protected facility.

Vendor and Service Personnel

Personnel servicing the facility will be required to enter from time to time. The type of personnel referred to may be telephone repairmen, air conditioning maintenance personnel, or vending-machine servicemen. They may enter on foot or may be authorized to take their vehicles inside.

The control of these individuals, as they enter and leave, should be similar to the control exercised over visitors, because these individuals are as much strangers to the facility as are visiting salesmen and others. The degree of security established to control the visitor will normally suffice for the vendor personnel.

Contractor Personnel Control

Since almost every industrial facility at one time or another will have some construction or renovation being performed by outside contractors, their control must be considered as briefly discussed earlier. The size and duration of the project and the number of personnel employed by the contractor and subcontractors would be determining factors in the amount of control exercised.

When construction of long duration is scheduled within the protected area, a temporary chain link barrier similar to that used to protect the entire facility may be erected. The barrier would be installed completely around the area where the construction is in progress, and temporary gates created in the perimeter barrier to lead directly into the construction area. The gates are normally manned by additional security personnel with special instructions for the identification and control of the contractor or personnel. If this is possible, little or no additional controls over contractor personnel may be needed, since this area would be fenced completely outside the protected property.

If it is uneconomical to secure the area as outlined above, control established for these personnel should be similar to the control established for visitor and vendor personnel at the facility. If the construction area is remotely located inside the protected area, control may be established through the deployment of a patrolling guard. It would be highly unlikely that construction personnel could effect surreptitious ingress and egress of the controlled buildings.

Contract Custodial Personnel

Personnel employed to clean certain areas of the facility should always come under security control. All such contract personnel should be required to sign in and out on a separate register and should be issued identification, because, in all probability, they will require access to more than one specific area.

Off-Schedule Employees

An *access roster* is merely a list of the names of individuals employed at a facility who are authorized by management to enter the plant during periods other than their regular working schedules. These rosters usually include maintenance personnel who may be required to return to the plant at times other than their regular shifts, and may even include some categories of company salesmen, laboratory technicians, or others who have need to be in the facility other than during their normal work schedule.

The access rosters are issued by plant management and should include the control number of each individual's identification media. The roster should be reviewed by management from time to time to insure that it is up-to-date, that personnel appearing on the roster are still employed, and are still authorized to enter the facility during odd hours.

Dismissed Employees

Employees who have been fired often create a problem in recovering the identification media issued. If the system is to remain effective, a method of recovery must be devised. Occasionally, the disgruntled employee who has been dismissed will claim a loss of the identification card. Usually, if the employee is required to sign a sworn statement to the effect that the card is lost it may help to deter his plans to use the card to attempt reentry.

Close liaison between the security force and the personnel manager should be established and maintained. The security force should be notified on a daily basis of the names of the individuals whose employment has been terminated and the cards not recovered to assist them in enforcing their responsibilities in personnel control.

Control of Materials

Identification buttons and paper stock used in the manufacture of the laminated identification card must be safeguarded to preclude compromise of the identification system. If the paper stock used to manufacture the laminated identification card has been stolen, the entire system is compromised. It must be discarded, and a new system established.

Checking Identification

Checking identification cards is usually the function of the security force. The guards stationed at control points must be furnished with adequate illumination in order to compare the photograph on each badge with the person who is wearing it. Guards stationed at control points must not become lackadaisical in enforcing and checking the identification system. These functions must not be allowed to become purely mechanical; this attitude would soon render the identification system ineffective.

Test attempts to penetrate the security plan at the control points, with the guards knowing that such attempts will be made, will assist in maintaining a high degree of alertness. This procedure is particularly effective if the security director arranges for complete strangers to make the attempt.

CHAPTER 10

Identification and Control of Vehicles

The identification and control of vehicles is directly related to the subject of identification and control of personnel, since the purpose of controlling both personnel and vehicles at points of entrance and exit to the protected area is to insure that only authorized persons and vehicles are permitted within.

Methods used to identify and control personnel only, were previously discussed. However, it must be remembered that at these points of entry and egress controlled by security personnel, a single guard may have the responsibility of identifying and controlling both the personnel and the vehicles passing through.

Identification of Employee-Owned Vehicles

Identification must be established if employees' vehicles are to be allowed entry to the property of any facility. Therefore, the vehicles must be provided with a means of identification. The simplest method of controlling and identifying the privately owned vehicles of employees is to issue an identification decal which is weatherproof and contains the company name and the control number. This decal should be issued to the employee at the same time his personal identification card is issued if he is going to be permitted to park his privately owned vehicle on company property. Ownership should be established through vehicle registration cards.

The vehicle identification should not be placed on the inside or outside of the windshield or rear window of the automobile. Many states prohibit placing decals on the front or rear windows of an automobile. Decals should not be placed on the left front window either inside or outside the vehicles since, during warm weather months, it may be necessary for the driver of the vehicle to roll his window up so the decal can be seen. Decals placed on these windows are soon damaged by raising and lowering the window.

Ideally, the decals should be placed on the left of the front bumper of the vehicle for quick identification by the security force since, in most instances, vehicles approaching gatehouses approach on the right side of the highway with the gatehouse in the center of or on the left side of the roadway to facilitate servicing the exiting vehicle. A decal on the left front bumper can easily be read as the vehicle approaches the protected area.

Decals may be placed on the forward-facing side of the rear view mirror where they would not obstruct vision but could easily be seen from the outside as the vehicle approached or as patrols are made of the parking area.

The greatest objection to placing the identification decals anywhere inside the vehicle is that in order to remove the decal, almost certainly the employee will have to be present to unlock the vehicle to permit the guard to remove and destroy the decal.

If the decal is placed on the outside of the automobile, the personnel office can notify the guard force that the employee is being dismissed. While his documents are being processed in the personnel office, the guard can move to the parking lot, locate the employee's vehicle, and effectively destroy the decal. This will eliminate the possibility of a disgruntled employee attempting to regain entry after his employment has been terminated.

The following control procedures should be considered in establishing the vehicle identification system.

1. The controls established to safeguard and issue identification cards should be followed as closely as possible.
2. The design of the decal should be intricate enough so it would be difficult to duplicate.

3. Each decal should be serially numbered, the numbers should be sufficiently large so they can easily be read from a distance, and consideration may be given to color-coding the decals to correspond with specific designated parking areas.
4. If the plant regularly works the night shift, consideration may be given to having the decals manufactured of a reflective material to facilitate rapid identification by the guard.

Restriction of Private Vehicle Parking

Emphasis is again placed on prohibiting parking within the same perimeter barrier that protects the plant proper. Again, this is most easily accomplished by fencing the parking lots away from the plant buildings. If private vehicles are allowed to park within the protected area (see Fig. 10-1), the security against pilferage is immediately reduced and other security measures are more difficult to enforce. The restriction of not allowing employees' private vehicles inside the protected area should also apply to supervisors and executives.

Visitors should be excluded from parking in any of the controlled areas within the plant site. Visitors' parking spaces should be as near as possible to the entrance. These parking spaces must be outside the perimeter barrier, since parking outside the perimeter barrier eliminates the need for extra regulatory controls.

Control of Trucks

Control of truck traffic is much more effective if the physical layout of the facility will permit trucks servicing the facility to use a separate vehicle entrance. This will not only greatly increase the efficiency of exercising control over registration and inspections of the vehicles, but will accomplish this without undue delay to other traffic. All trucks authorized to enter the property should be registered on a preprinted vehicle log and should contain at least the following basic information:

1. The name of the company to which the truck belongs
2. The name of the truck driver and any assistants who will be authorized to enter the protected area with him.

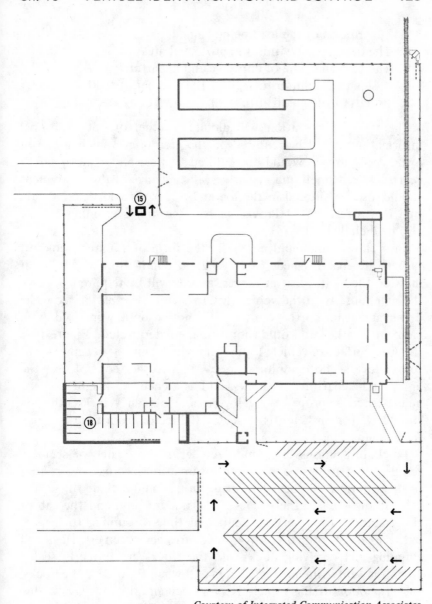

Courtesy of Integrated Communication Associates

Fig. 10-1. Identification and control of vehicles is the primary function of the truck gate guard (15). When office personnel are privileged to park in areas other than the main employee parking lot (18), the means to identify and control vehicles using these areas must also be provided.

3. Description of the load being hauled
4. The destination within the protected area
5. The date and time of entrance and departure
6. A separate column to register the number of the truck trailer
7. Initials of the guard authorizing ingress or egress.

The type of operation will dictate whether or not additional information would be required on the preprinted vehicle log. For example, if drivers would be required to wear any type of safety gear such as helmets and eyeglasses, the issuance of this equipment should also be entered in the log and a space provided to indicate that it was recovered. Provisions for recording seal numbers may also be included.

The vehicle log should not be in the form of a bound book but rather on separate sheets so that upon completion of a sheet it may be filed. In most instances, guards will be required to leave the gatehouse with the vehicle log to secure information from the driver and from on the vehicle. If a bound book were used in inclement weather, it would soon become deteriorated. Figure 10-2 shows an effective vehicle log system which employs a camera unit to produce a photographic record. As a further control over the entry and departure of trucks, and as a check to insure that the guard force is properly logging in and out all vehicles, the log should be forwarded to the security director from time to time, for examination.

Truck inspections are conducted when the degree of security requires that trucks be inspected for unauthorized passengers and/ or unauthorized cargoes. This inspection should include the underside of the chassis, the inside of the trailer body, and the cab of the tractor. It is very difficult and time-consuming to inspect entire loads. If proper control procedures have been established at shipping and receiving docks and the trucks are properly sealed when they enter the protected area and when they leave the docks, the inspection of the vehicle is simplified and the process speeded up considerably.

Some techniques which have been successfully used to control the movement of trucks to and from the shipping and receiving

Courtesy of Scan-O-Scope

Fig. 10-2. Vehicular control through use of this camera (top) can be auto-matically or manually controlled. The documentation (bottom) includes identification of the vehicle, the bill of lading, the driver's license, and the exact time the vehicle entered or left the protected area.

docks when the dock areas are not under the direct observation of the guards are as follows:

1. If the requirement and degree of security is high enough, the use of closed-circuit television surveillance cameras along the route may be justified. This system would have to be backed up by a guard to monitor the cameras and also by additional lighting if truck movements are authorized after dark. It may also require an additional guard or communication to an already established outside roving patrol to facilitate dispatch of a guard should the truck deviate from the prescribed route.

2. Timing trucks on trips from the gate to the dock area, and return, can also be effectively used. To establish this system, a gate pass is issued to the truck driver by the guard at the entrance gate with a description of the truck, the name of the driver, and the exact time the truck departed the entrance gate. The driver must be required to give this pass to the shipping or receiving dock supervisor or clerk. This individual should immediately examine the pass and determine whether or not the truck has deviated from the prescribed route, for he will have previous knowledge of the exact time consumed to negotiate the route. The reverse of this procedure is initiated when the truck leaves the shipping or receiving dock and returns to the vehicle exit point.

   The times to and from designated destinations within the facility should be predetermined by actually traveling the prescribed route at the prescribed rate of speed. It should be remembered that the prescribed route must be adequately marked, so that truck drivers unfamiliar with the area do not innocently deviate from it.

3. In some instances, it may be possible to ditch, fence, or install guard rails on the sides of the main thoroughfares and/or primary roadways. When secondary roadways create junctions off the primary roads, these roadways may be controlled by stationing guards or establishing permanent or temporary barriers to restrict unauthorized movement off the main arteries.

4. Where the protected area is very large in size and multiple roadways exist and are used constantly, the installation of

additional directional signs may be required or some routes may be color-coded by placing signs of the appropriate color along that particular roadway. This system is used at most of the large airports to designate routes to specific terminal buildings, parking lots, and airline terminals.

Utility Vehicles

Control of utility and vendor vehicles is necessary, since some of these vehicles will have to move within the protected area in the normal servicing of the facility. These may be the vendor service-men's vehicles, utility company vehicles, or local delivery vehicles. Basically, the vehicles should be controlled by registering the ve-hicle and the driver and identifying his vehicle in a manner similar to that used to control other trucks.

Further control over these vehicles can be established by issuing a card approximately six inches square containing a large control number and the identification of the facility. The driver should be required to display this identification card in the left front of his windshield while he is within the protected area. This procedure will assist in the control and policing of the vehicle internally and the number is used to identify the vehicle merely by calling the gatehouse. Cards may also be color-coded to specific areas, if the security requirements so dictate.

Sealing trucks moving in and out of a facility is an important part of the overall control program as described in Chapter 7. In some instances, outside truck drivers object to having their trucks sealed because they have additional deliveries or stops to make other than the facility they are currently servicing.

Past experience has indicated these objections will often involve complaints from the truck drivers' union. Should this situation prevail the trucks should still be sealed when they depart the dock area. The guard at the point of exit should check the seal number when the vehicle arrives, record the number, and if necessary, remove the seal. The reverse of this procedure is used when trucks are sealed at the time they enter the premises to insure pilferage is not being committed between the point of entry and the desti-nation within the protected area.

CHAPTER 11

Protective Alarms and Systems

In previous chapters four main lines of defense or protection were discussed: the perimeter barrier; area security; security of the peripheral walls of the buildings; and security applied to the numerous areas within buildings.

If alarm systems are used in conjunction with these main lines of protection, the efficiency is greatly increased, and the security plan is no longer entirely dependent on the strictly physical barriers.

In general, seldom is a single aid to security considered sufficient to provide absolute protection of an area or premises. It is more desirable to provide a series of aids to complete the whole protection plan. Alarms will normally be needed only in those areas with high priority on the established criticality list, and the proper application of alarm systems will assist in materially reducing vulnerability. The facility site plan, Fig. 11-1, shows areas where different types of alarms may be needed.

The origin of alarm systems to protect life and property dates back at least a century. Over many years a variety of mechanical devices has been contrived in an attempt to detect intrusion.

The manufacture of various types of alarms today is a lucrative business. Many gadgets purporting to be alarms are offered to the public. The great majority of these alarms are very limited in their capabilities and, because of inferior manufacture, many false signals result. One authoritative source estimates that over a million dollars is being spent annually for worthless alarm devices.

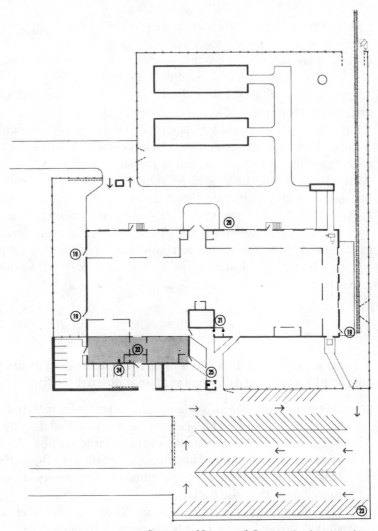

Courtesy of Integrated Communication Associates

Fig. 11-1. The protective alarm system may consist of local door alarms (19) on emergency exits, photoelectric alarms (20) inside the peripheral walls of the dock area, electromagnetic door switches (21), ultrasonic or RF motion detectors (22), taut-wire perimeter protection (23), and vibration or capacitance protection of safes and files (24). A proprietary system would include monitoring all but local alarms, at the main guardhouse (25).

However, in recent years alarms have been improved, and those now offered by major manufacturers and approved by Underwriters' Laboratories, Inc., are reliable and can be depended upon to furnish a fairly high degree of effectiveness.

This chapter will deal primarily with a general explanation of how alarms are monitored, the alarms available to assist in establishing the security program, and a brief explanation on how they function. It should be kept in mind that the alarm must be selected because of a specific need to establish a higher degree of security in a particular area. If the degree of security of a particular office or building requires only that the door be locked with a substantial locking device, then to install any type of anti-intrusion alarm device would be uneconomical. Therefore, when considering the installation of alarms one must consider whether the degree of security being established is necessary to successfully complete protection.

Alarm Systems

There are basically two types of alarm systems used in formulating the physical security program of an industrial facility:

1. *Anti-intrusion alarm systems* are commonly referred to as "burglar alarms." These systems are installed for the sole purpose of detecting unauthorized intrusion when the alarm systems are activated. They may be installed on the perimeter barrier, the periphery of the building, or merely at certain positions within the building.

2. *Fire protection alarm systems* are those alarms which are installed throughout a facility to protect it against fire. These alarms range from the water flow alarm to the heat sensors, smoke detectors, and fire-door monitors. The fire protection alarm systems will be discussed in Chapter 13.

Another type of alarm, the process alarm, is a device installed to monitor a particular process. Although process alarms may be monitored by the security force, they are not considered in the physical security study.

Monitoring Systems

There are four different methods of monitoring the various types of anti-intrusion and fire protection alarms.

1. *Central station monitors* are located in urban areas and may be miles from the protected property. A pair of ordinary telephone lines is used to transmit the signal from the alarm to the monitor panel. Operators monitoring the panel call the fire department or the local police department when an alarm is received.   •

   Some central station companies will have personnel available who can be dispatched to conduct investigations. However, alarms monitored at central stations could require a considerable amount of time before investigation can be made because of the large distances involved.

2. *Proprietary station monitors* are installed on the property being protected. Security forces on duty at the facility monitor the alarms and respond when an alarm is activated. Since the alarm is monitored on the property, the distance between the monitoring panel and the alarm is considerably less, and guards can react to the alarm immediately. Obviously, this is the most desirable method of monitoring any type of alarm.

3. *Local alarm monitoring* is done in the immediate vicinity of the alarm. No monitoring panel is usually necessary, because an audio device is an integral part of the alarm itself. But if the installation being protected is very large, a guard in one part of the building may not be able to hear an alarm in another section. If employees are present in the facility, the local alarm provides adequate notification. The sound emitted by the alarm is usually from a loud bell, horn, or siren. The local alarm system is excellent for fire protection, because when the alarm is activated all those in the immediate area of the alarm are warned. However, the signal must also be transmitted to the firefighting department.

4. *Auxiliary or remote monitoring* is accomplished by locating the monitoring panel in the local fire department or police department servicing the protected property. When this system of monitoring is used, a local alarm is also sounded. Since

the fire or police department is located outside the protected areas, leased pairs of telephone lines must be used to transmit the signal. When a signal is received at either the fire or police department, it is charged with the responsibility of responding.

Alarm Types

One can think of an alarm, whether it is for anti-intrusion or fire protection, merely as a triggering device. This triggering device is activated when certain conditions occur. The types of alarms available are many and varied. It must be remembered that the type of protection to be developed is dependent upon the type of alarm selected.

Anti-intrusion alarms have a definite place in the security program—if the proper type of alarm is selected, if it is properly installed, and if it is correctly adjusted. There are several types of anti-intrusion alarms available.

Space or motion detection alarms are classified into two categories—1. radio frequency and 2. ultrasonic. These alarms are designed to detect motion within the protected area, and, as the name implies, space rather than an object is being protected. However, objects within the protected space obviously are included.

1. Radio frequency alarms operate by emitting radio frequency waves throughout the protected area between a transmitting and receiving antenna. Once properly adjusted, the alarm stays in balance as long as these radio frequency waves are not disturbed. If the protected area is entered, it upsets the pattern of the waves and an alarm is triggered. This alarm's use is limited, because the radio frequency waves will penetrate the walls of the protected area unless they are lined with metal. If the alarm is not properly tuned to the area, the radio frequency waves penetrating the walls could be unbalanced by a person walking outside the protected area, creating a false alarm.

2. Ultrasonic alarms also consist of a transmitter and receiver. The transmitter emits ultrasonic sound which is tuned so high that the human ear cannot detect it. When the alarm is

properly adjusted, sound waves are emitted from the transmitter and are picked up by the receiver. As long as the sound waves are not disturbed, the alarm will not be triggered. Anyone entering the protected area interrupts the sound waves pattern, throwing the alarm out of balance and triggering an alarm. (See Fig. 11-2.)

This type of alarm also has certain limitations. Ambient noises or noises created outside the protected area which also emit sound waves could upset the sound wave pattern, causing a false alarm. Air conditioning or heating vents pouring air into a room protected by the ultrasonic alarm may also trigger a false alarm.

Courtesy of Honeywell, Inc.

Fig. 11-2. The ultrasonic motion detector fills a predetermined amount of space with ultrasonic waves. Illegal entry upsets the balance of the device which then signals the condition at the monitor.

Audio alarms are sometimes referred to as "sound" or "sonic" alarms. Generally described, they are merely microphones placed in the protected area which will transmit all sound occurring in that area to a receiver which must be monitored. Typical equipment is shown in Fig. 11-3. These alarms can remain on at all times, or they can be adjusted so a sound loud enough will activate the system and a buzzer or light will occur where the alarm is being monitored. The guard will then turn on the receiver and listen to the sounds. In recent years, this alarm has undergone sophistication, and an additional piece of equipment referred to as a "sound discriminator" is now wired into the system. The alarm is adjusted so that ordinary noises in the protected area will not activate the alarm, but unusual noises will trigger it.

Courtesy of Honeywell, Inc.

Fig. 11-3. Audio detection devices consist of microphones of varied sizes and shapes and a control panel. Sounds activate a signal to alert the guard at the monitor location. The small device in the foreground is a vibration detector. Used on walls or roof supports it signals attack to the remote monitor operator.

Photoelectric alarms utilize a beam of ultraviolet light, which cannot be seen with the naked eye. The alarm consists of a transmitter and a receiver. The transmitter emits the beam of ultraviolet light to the receiver. As long as this beam of light is not broken by an object moving through it, the alarm stays in balance. If the beam is broken, the alarm is triggered. Because a beam of light is used, the protection furnished is always on a straight line. However, mirrors can be used to direct the beam of light in another straight line direction, but the beam of light must always terminate at the receiver. The size of the photoelectric alarm will depend upon the length of the protective beam of light. Photoelectric alarms have a maximum range of from 25 up to 600 feet.

These alarms are often installed inside a window, door, or wall, so that anyone violating the protected area will interrupt the beam of light, thus triggering the alarm.

Capacitance alarms are usually used for "object protection," protecting a particular file cabinet, safe, or other metal object. The alarm operates on the principle that when properly installed and activated, it creates an electromagnetic field around the protected object. The object must be metal and must be insulated in such a way that it is not grounded in order to establish the electromagnetic field. The alarm is usually adjusted so that the magnetic field produced is emitted only ten to twelve inches from the protected object, permitting such people as cleaning personnel and employees working late to move about the room where the object is stored without activating the alarm. However, when an individual attempts to touch the protected object or otherwise move his body within the electromagnetic field, the electric capacitance is changed and an alarm is triggered.

Vibration detection alarms are also used for object protection. A series of these alarms may be installed along a wall or window. If the wall or window is attacked an alarm will be triggered. These alarms operate as a result of vibration. Therefore, if the object being protected is attacked, the vibrations will trigger the alarm. As with all other alarms, the vibration detectors can be tuned to different degrees of sensitivity. The type of protection being established will determine how sensitively the alarm is set. A detector is seen in the foreground of Fig. 11-3.

Electromagnetic devices are commonly used in door protection. They consist merely of two magnets, one installed on the door, the other on the frame. When the alarm is activated, a magnetic field is created between the two magnets. If the door is opened, the magnetic field is broken and an alarm goes off. An installation is shown in Fig. 11-4.

Courtesy of Honeywell, Inc.

Fig. 11-4. Magnetic control switches are effectively used to monitor doors in the security system. Signals are relayed to the monitor when doors are opened during the period the system is activated.

Electromechanical devices are nothing more than ordinary mechanical switches. They are usually held in place or in the open position by simple spring devices. As an example, if one were installed on a door, the plunger would be pushed in when the door was closed, making no electrical contact. If the door were opened, the plunger being released by the door would be pushed outward through the action of the spring, and an electrical contact in the switch would cause the alarm to be activated.

Pressure devices operated on the principle that when a certain amount of pressure is applied to the device, electrical contact is made and the alarm triggered. Simply stated, although they are manufactured in numerous forms, this type of device is the same type we see when entering stores or other public buildings. Pressure on a rubber mat causes the doors to automatically open. In security applications, these devices are camouflaged and would trigger an alarm to indicate unauthorized intrusion. (See Fig. 11-5.)

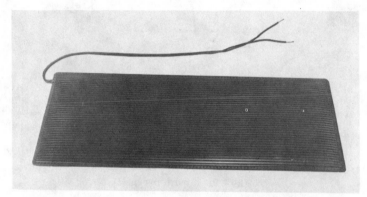

Courtesy of Honeywell, Inc.

Fig. 11-5. A floor mat pressure switch, normally used to open doors electronically, can also be incorporated into the security plan. Pressure on the hidden switch can signal unauthorized intrusion.

Foil tape devices, with which almost everyone is familiar, are found on numerous storefront windows. The alarm operates on a very simple system of continuous electrical circuit using a very fine wire embedded in the tape. When the tape is broken by smashing the window, the electrical circuit is broken, and an alarm is triggered.

Taut wire detectors are installed near the top of the perimeter barrier and offer additional protection in this first line of protection. One unit including a control panel can be used to protect up to 1,000 linear feet of perimeter. A sufficient amount of detection wire and additional panels can be installed as required. (See Fig. 11-6.)

The taut wire is held in tension by a small weight at one end, and at the other end a spring mechanism within the taut-wire control

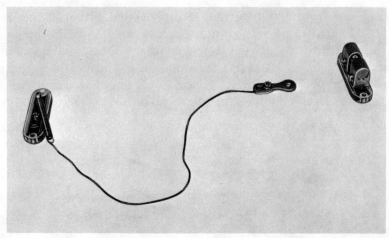

Courtesy of Honeywell, Inc.

Fig. 11-6. The trip wire contacts can also be used as a taut-wire alarm signal on the perimeter barrier. Pressure or release of pressure on the device activates the intrusion signal.

panel.   The panel contains a two-way switch so that either a relaxation of the tension or an increase in the tension will trigger the alarm.

Other outdoor perimeter protection equipment, mostly in the experimental stage at the time of this writing, includes an *electronic fence* consisting of nine wires, or three levels of three wires each, installed along the perimeter inside the chain link fence or other barrier.   The surface beneath the electronic fence must be paved to prevent swaying weeds from causing a false alarm.   The fence creates a field of electrical energy which, when disturbed, is placed out of balance, triggering an alarm.

Some manufacturers are experimenting with *radio frequency systems* which, when developed, would also increase the protection of the installed perimeter barrier. These devices consist of a transmitting and receiving antenna using radio frequency waves. When the antenna load is changed by a person entering the invisible beam, an alarm signal is actuated.

Protecting Power Supplies

There are probably numerous readers who have already decided that any of the types of alarms described above can easily be defeated merely by interrupting the power supply. Alarms would have little value in the security program if this were true. All alarms do require electrical wiring between the alarm or triggering device and the monitoring panel, and rely upon electrical power for their operation. Obviously, if this line can be found and attacked, the alarm can be deactivated. To insure that the alarm is not deactivated by this means, the electrical lines are supervised by very low voltage constantly surging through them. If the line is cut or the voltage otherwise changed, an alarm is automatically triggered.

Standby power is provided for, should a power outage occur. It is supplied in the form of dry cell or wet storage batteries. The standby power is sufficient to provide four to twelve hours' operation and can transmit the alarm signal to the monitoring panel when the protected area or object has been violated.

CHAPTER 12

Closed-Circuit Television
Surveillance Systems

It was only a few short years ago that the closed-circuit television (CCTV) industry was still in its infancy. In recent years great strides have been made in perfecting the television camera, the monitor, and all its related equipment.

Closed-circuit television surveillance systems have been successfully incorporated into industrial plant protection, and today play an important role.

This chapter is devoted to a discussion on the application of such systems and the important role they can play in a properly engineered protection plan. It does not contain highly technical data. The actual installation of closed-circuit television should be assigned to the experts in the industry. However, the security director should have a working knowledge of what these systems consist of, where they may be effectively applied, and, generally, the capabilities and limitations of these systems.

CCTV Costs

The proper application of closed-circuit television in the protection plan can very often dramatically reduce the manpower costs of the program. One must keep in mind that even though an immediate capital expenditure may be required, the reduction in manpower over a period of time will usually result in tremendous labor cost savings. Leasing plans which require little capital outlay are offered by most television suppliers.

The security director is always confronted with the intitial cost of any equipment he seeks to use in the protection plan. He must justify each expenditure, more so probably than any other expenditure in an industrial plant. This is probably because accounting figures for security purposes always appear on the "payable" side of the ledger, whereas, unfortunately, security force receipts cannot and will never appear on the "receivable" side of the ledger.

Where the above conditions apply in any particular instance, initial justification should be directed toward the actual reduction in manpower which the closed-circuit television system would permit. Keep in mind that fullest economy from such a system can be realized, if each individual camera is used on a twenty-four-hour-per-day basis wherever possible. Proper application of this type of surveillance will normally result in total and constant surveillance of a given area or areas.

Secondly, consideration must be given to the method which will be employed to monitor the camera. The effectiveness of any closed-circuit television surveillance system in the security plan will almost surely depend upon its being monitored by a member of the security force.

If a twenty-four-hour fixed post is or will be established, this would normally be the point where the system is monitored and there should be no additional labor cost involved in monitoring. It is quite obvious that to establish an additional post merely to monitor the surveillance system could conceivably increase the cost of security rather than cut it down.

Thirdly, provision must be made to respond and initiate remedial action if unfavorable conditions in the surveyed area are detected.

If these three conditions can effectively be met without increasing security manpower hours, most surely the addition of closed-circuit television to the overall security plan can be justified. Perhaps the additional security obtained from a surveillance system by itself will justify not only the expenditure for equipment, but for additional security guards as well.

CCTV Security Applications

There are numerous areas in an industrial facility where closed-circuit television surveillance can be applied practically (see Fig. 12-1). The major areas are discussed below; however, the discussion is not intended to limit the use of this type of surveillance. The complete study and analysis of each individual facility could possibly reveal other areas where such systems could be employed effectively and economically.

Closed-circuit television can be effectively used as a surveillance method at electrically operated, remotely controlled vehicular or personnel perimeter gates. This possibility is particularly suited to gate control application where several gates are being used intermittently. A single security guard stationed at the most active point of ingress and egress could effectively control several gates. This application would be far more economical than an attempt to control additional gates with security guards and most certainly would enhance the security at these gates, if they had been left open for specified periods of time, without control, in the past. A simple system is shown diagrammatically in Fig. 12-2.

Surveillance of *perimeter fences*, particularly in remote areas, can be accomplished far more efficiently and economically than by patrolling security guards. A system suitable for this purpose is shown diagrammatically in Fig. 12-3. Greater security in this application is realized because the area will be constantly under observation as opposed to being under observation only during those periods that the security guard is physically present in the area.

In *shipping and receiving operations*, particularly when operations cover large areas and the material being shipped or received can be readily pilfered by an individual, closed-circuit television is excellent. Closed-circuit television surveillance of dock areas can be used in conjunction with security guards physically patrolling the areas. The surveillance system would be monitored at a fixed guard post, probably at a remote location, and unfavorable incidents observed by the monitor would be relayed via radio communications to the patrolling security guard, who would respond immediately.

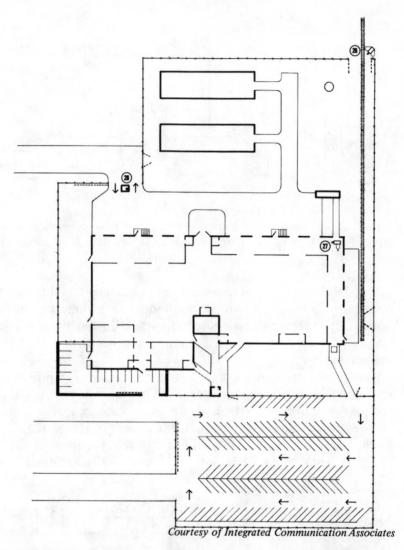

Courtesy of Integrated Communication Associates

Fig. 12-1. A closed-circuit television system might include surveillance of remotely located, electronically operated gates (26), or remote sections of the perimeter barrier. Observation of dock operations (27) will deter pilferage and reduce damage as well as increase safety in industrial vehicle operation. The surveillance systems may be monitored at the truck gate (28), or elsewhere.

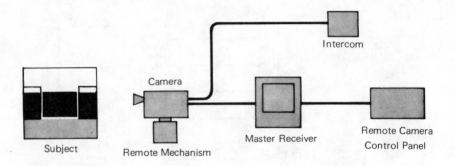

Intercom

Camera

Master Receiver

Remote Camera Control Panel

Subject

Remote Mechanism

Fig. 12-2. The closed-circuit television camera and intercommunication system are used to establish effective control at remote gate locations with low-density traffic.

The deterrent value of the television equipment is difficult to measure. However, it would be prudent to assume that many potential pilferers or thieves have been deterred from illegal activity simply because they had no assurance whether they were being observed or not.

Warehouses are particularly suited to establishing protection through the application of closed-circuit television. A system recently developed is particularly suited to this need. The system referred to operates on a pneumatic "rail" and is, therefore, silent. Cameras travel at speeds from 40 to 250 feet per minute, with a stopping distance of from 4 to 8 inches. Rails or "track" are

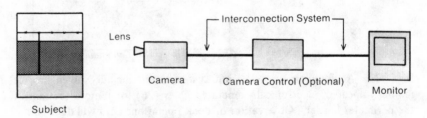

Interconnection System

Lens

Camera   Camera Control (Optional)

Monitor

Subject

Fig. 12-3. A simple closed-circuit television surveillance system may employ only one camera and a monitor. The camera control may or may not be utilized, depending upon specific application.

tailored for each individual application. From one to four cameras can be monitored in a cluster to afford observation in four directions at once, using four monitors. This system is said to cover 6 times the area observed by 10 fixed cameras using only one camera with pan and tilt capabilities. (See Fig. 12-4.)

Courtesy of OHM Manufacturing Corp.

Fig. 12-4. Closed-circuit television mounted on a rail, pneumatically driven and remotely controlled, is said to cover ten times as much area as a fixed camera. This application is particularly suited to surveillance of warehouse areas.

There may be numerous highly critical and vulnerable installations, areas, or materials at an industrial facility where closed-circuit television surveillance systems could be effectively employed. As an example, approaches to a particularly critical and vulnerable installation could be kept under constant observation. In warehousing operations, effective use may entail observation of critical supply storage areas or packaging rooms where the supplies are being prepared for shipment.

The application of closed-circuit television in the security field is really limited only by the ingenuity of the security director. The completeness and detail of the physical study and ultimate analysis of the findings will surely provide some argument for its use.

Recording Impressions

The use of video tape to record what the camera sees has un-limited applications. Consider, for a moment, employing a hidden camera to observe an area from six o'clock in the evening until six o'clock in the morning. No one monitors the camera, because video tape is being used. The following morning, in about one hour, all that the camera saw during the twelve-hour night can be reviewed.

Elements of the CCTV System

Closed-circuit television systems have certain limitations. These limitations may sometimes be the result of poor visibility because of inadequate protective lighting or inclement weather conditions. Television equipment is available to enable better observation of an area from remote locations with much better clarity than could be seen by the human eye. If the security analysis indicates prob-able use of closed-circuit television in the overall protection plan, local reputable dealers should be contacted and actual light meas-urements and/or actual sightings should be made to determine the type of equipment required to accomplish what is desired.

The closed-circuit television system's function involves five pri-mary elements. These elements are as follows:

1. The *subject* is the object being observed, whether it is station-ary, moving, or the result of a temporary breach in security. It might be, for example, a remotely located gate being opened by electronic means from the guard station.

2. The *lens* of the camera will determine, generally, the size and depth of the area which can be seen by the camera. There are numerous different lenses available which can be applied to practically any condition with extreme effectiveness. The selection of the proper lens is extremely important in securing the desired function of each individual camera. These lenses can be preset manually or can be electrically controlled from the remote monitoring location. The "zoom" lens is effec-tive when used in combination area and point surveillance. (See Fig. 12-5.)

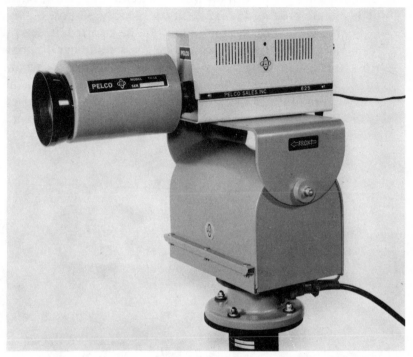

Courtesy of Pelco Sales, Inc.

Fig. 12-5. The zoom lens operated manually is effective in close observation of remote gates, doors, or personnel control points.

3. The *interconnecting system* consists of the coaxial cable link-ing the camera to the camera control and the monitor. If extreme distances between the camera and the monitor are involved, an image booster unit may have to be installed in order to obtain the desired image. The booster units are inexpensive and are usually installed if the distance between the camera and monitor is in excess of 1,000 feet.

4. The *monitor*, or viewing device, is much the same as the home television receiver. Television pictures or images are produced by a number of lines which are defined as resolution in lines in relation to the height of the picture. Most of the industrial television cameras will produce pictures with 400 lines or

more horizontal resolution when adequate light is provided. The picture produced is excellent for security purposes.

Economy in installation can often be realized if several cameras can be effectively used with a lesser number of monitors. This can be obtained by having a video switcher installed between the cameras and the monitors to switch the image seen on the monitor between the cameras at given periods of time and in a specific sequence. A control panel is shown in Fig. 12-6.

Courtesy of Pelco Sales, Inc.

Fig. 12-6. Control panel for a single-camera, closed-circuit television system.

5. *Accessories* will be required in most instances when the camera is installed at an industrial complex.

If the camera is installed outside, perhaps for perimeter barrier surveillance, it must be housed in a weatherproof housing for the protection of the camera.

More often than not, the weatherproof housing will have to be equipped with a windshield wiper and may require a heating element if weather conditions are severe. Various types of available housings include explosion-proof housings, dustproof housings, and housings capable of withstanding high heat. (See Fig. 12-7.)

In almost all security applications, the television camera will have to be equipped with a pan and tilt device. The device permits the operator and monitor to move the camera on a vertical and horizontal plane. This can be accomplished either manually or automatically.

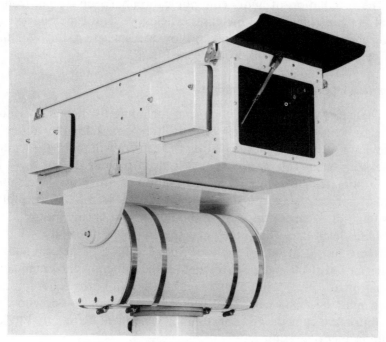

Courtesy of Pelco Sales, Inc.

Fig. 12-7. The weatherproof closed-circuit television camera includes a windshield wiper and a variety of accessories.

Personnel Control

Closed-circuit television cameras can also be effectively used for the identification and control of personnel. For example, an unauthorized point of ingress and egress for certain categories of personnel may be located far from the primary employee control point. Two closed-circuit television cameras are installed, and the opening through which personnel will pass is controlled by a turnstile which is electrically operated and remotely controlled by the security guard at the primary point of ingress and egress.

The two cameras are mounted so that the uppermost camera is viewing the face of the individual and the lower camera transmits a picture of the issued identification card which is placed near the camera lens. The security guard makes his identification and releases the turnstile control to permit entry and exit of personnel.

The installation of audio communication between the controller and the controlled point increases efficiency. This type of application should be considered only in those areas where a small volume of pedestrian traffic will occur. (See Fig. 12-2.)

Camera Operation and Location

The exact location of the camera, including the height at which it is to be installed, can most effectively be determined by actually moving a person into the position the camera will take. Generally speaking, what this person sees at a given time of day or night is approximately the amount of observation afforded by the camera unless, of course, very low light cameras are used. Before actual installation is made, the camera should be temporarily mounted to double-check that it is in the most advantageous position.

After the location of each individual camera is determined, a decision must be made as to whether the camera will be manually or automatically operated, or a combination of both. A determination must be made where the camera will be monitored, who will monitor the camera, and the training which will be required to insure proper operation.

Camera maintenance must next be considered as extremely important in security operations since, in all probability, observation of the area being surveyed by the camera will not come under observation by security guards other than those responsible for monitoring. Camera maintenance will probably have to be performed by the installer through a negotiated maintenance service contract agreed upon at the time of installation. It should be expressly understood with the maintenance contractor that when service is required and called for he must respond immediately. To insure that the service is complete, the adequacy of the supply of spare parts should be determined by the security director through a visit to the contractor's place of business.

CHAPTER 13

Fire Protection and Prevention

The protection of the industrial facility against fire is among the most important functions of the security force; in many instances, it *is* the most important function. Not only is the security force charged with the responsibility of inspecting housekeeping of the entire facility and reporting fire hazards, but, because they are often in the facility when few or no other individuals are present, they are charged with the responsibility of initiating rapid communications with the servicing fire department and designated plant personnel. They must also be capable of immediately attacking a fire with the means at their disposal.

The purpose of this chapter is twofold. It is intended both as a review for security personnel who have been exposed to some degree of formal training in the subject and as a training aid for the security director responsible for fire protection and prevention training of his security force. The information is outlined in a logical sequence to cover the subject of fire protection and prevention as it applies to most industrial facilities. However, establishment of a fully effective fire protection and prevention program for an industrial facility may require the services of a professional fire prevention specialist, and this chapter is only intended to introduce the basic principles.

The direct savings or loss prevention value realized through the implementation of an effective fire protection and prevention program are extremely difficult to estimate just as it is no easy matter to estimate any direct savings resulting from the protection

plan as a whole. What is more important than attempting to estimate direct savings is to provide adequate training to avoid the
losses that can occur through fire. In the past, proper training of
responsible people has often been neglected and the tools to protect against the potential of fire were not available or were insuffi-
cient for the job.

Elements Necessary to Start a Fire

Several elements must be present in order for a fire to start
(see Fig. 13-1.)

Fig. 13-1. The three elements that must be present for a fire to occur are:
heat or friction, some fuel, and oxygen.

1. *Fuel,* in whatever form, must always be present, because there
   must be some type of combustible material for the fire to
   consume.
2. *Oxygen* in sufficient quantities for the fire to feed upon is
   necessary, and the supply of oxygen must be present in vary-
   ing degrees of volume if the fire is to be sustained. The larger
   the fire, the greater the quantity of oxygen required to sustain it.
3. *Heat or friction* is the third element required to start a fire.
   Ignition sources can range from unseen spontaneous combustion to the simple ignited match.

Four Stages of Fire

Fires generally go through four stages. In the *incipient stage,* no visible smoke, flame, or any significant heat is yet developed. However, a condition already exists which generates a significant amount of combustion particles. These particles have mass but are too minute in size to be visible to the human eye. The particles behave according to gas laws and rise quickly. This stage usually develops over an extended period of time, lasting minutes or even hours.

The *smoldering stage*, or smoke stage, as it is often called, occurs as the fire condition develops and the quantity of combustion particles increases to the point where their collective mass now becomes visible. During this stage, there is still no flame or significant heat.

In the *flame stage*, the fire condition has developed to the point where ignition actually occurs. Infrared energy is now given off from the flames. The level of smoke will usually decrease in the flame stage and the amount of heat developed will increase.

In the *heat stage*, or conflagration stage, tremendous amounts of heat, flame, smoke, and toxic gases are produced. This stage develops very quickly and usually follows the flame stage in only seconds.

Classes of Fire

The types of combustible materials consumed by fire have been classified into four categories, primarily as a means of identifying the particular class of fire with a specific type of extinguishant that is used to put it out. The reaction of any specific extinguishant on the fire depends upon the class of material being consumed. If the extinguisher is to be effective it must remove at least one of the elements that is necessary to sustain the fire.

Each of the four classifications of fire has been assigned a color-coded symbol (see Fig. 13-2) to identify the type of extinguisher to be used in the attack upon that particular class of fire.

Class A fires (green triangle) are those which occur in ordinary combustible materials, such as paper, wood, cloth, and grass. These fires are best extinguished by quenching them with water. Water

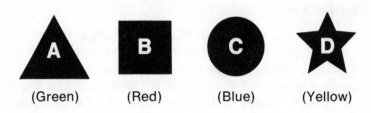

(Green)   (Red)   (Blue)   (Yellow)

Fig. 13-2. Fires are classified into four categories. Guards must be able to identify fire classifications with the type of extinguishant most suitable to put out the fire. Classification symbols of a specific color and shape are assigned, as shown.

lowers the temperature of the burning mass below the kindling point necessary for continued fire action to be sustained. Extinguishers other than water may be successfully employed to put out this class of fire; however, water produces the most immediately effective results.

Class B fires (red square) are those occurring in flammable liquids, such as gasolines, oils, paints, and cooking fats. Burning liquids require a smothering action to extinguish the fire. Extinguishers generating foam and dry chemical extinguishants are the most effective in subduing the Class B fire. Water may be used if nozzles and water pressure will create a "fog" by breaking the stream of water into minute droplets.

Class C fires (blue circle) occur in live electrical equipment, such as in motors, switches, and appliances. A nonconducting extinguishing agent is required to fight Class C fires; therefore, carbon dioxide or general- or special-purpose dry chemical extinguishers are the most suitable. A vaporizing liquid agent is also available. The reaction of this extinguishant results in the rapid vaporization of the liquid as it comes in contact with the burning material, creating a smothering effect.

Class D fires (yellow star) occur in flammable metals. These are metals such as magnesium and certain other special metals which will burn when chemical changes are in correct proportion. Special-purpose dry chemicals, powdered graphite, and other specially prepared powder agents are used in successfully extinguishing the Class D fire.

Portable Hand Extinguishers

The portable hand extinguisher is intended for use against small fires, usually those confined to a restricted area. Hand extinguishers cannot be expected to take the place of hoses capable of delivering large volumes of water in the fire attack or to replace the automatic sprinkler system. On the contrary, a great percentage of disastrous fires occur because the person discovering the fire attempts to extinguish it before the alarm is given.

If fire extinguishers are to be effectively used, they must be chosen on the basis of their suitability to extinguish that particular class of fire. Security guards must be able to effectively use extinguishers when fire emergencies arise, and they must be trained and thoroughly familiar with the type of extinguishant each type of fire extinguisher contains (see Fig. 13-3.) They must also realize that the discharge time of the portable hand extinguishers is usually a minute or less, and the quantity of extinguishant available in any one extinguisher is very small.

In industrial facilities where most employees in any given area are women, various types of portable hand extinguishers may be grouped on movable carts but if a fire should occur in the particular area protected, it probably still will be necessary to assign male employees to handle the carts and equipment because of the bulk and weight. (See Figs. 13-4 and 13-5.)

Placement of Portable Hand Extinguishers

In determining the specific type and number of fire extinguishers which are required and should be made available in any given area, readers should refer to the technical publications listed in the bibliography, or confer with their insurance agent.

If a survey of the numbers, types, and locations of the portable hand extinguishers has not been conducted in the recent past, a survey of the entire facility should be completed. The survey should include inspecting and testing each extinguisher, determining if the type of extinguisher at each location is suited for the class of fire most likely to occur, and whether the number and size of the extinguishers presently available comply with fire protection requirements.

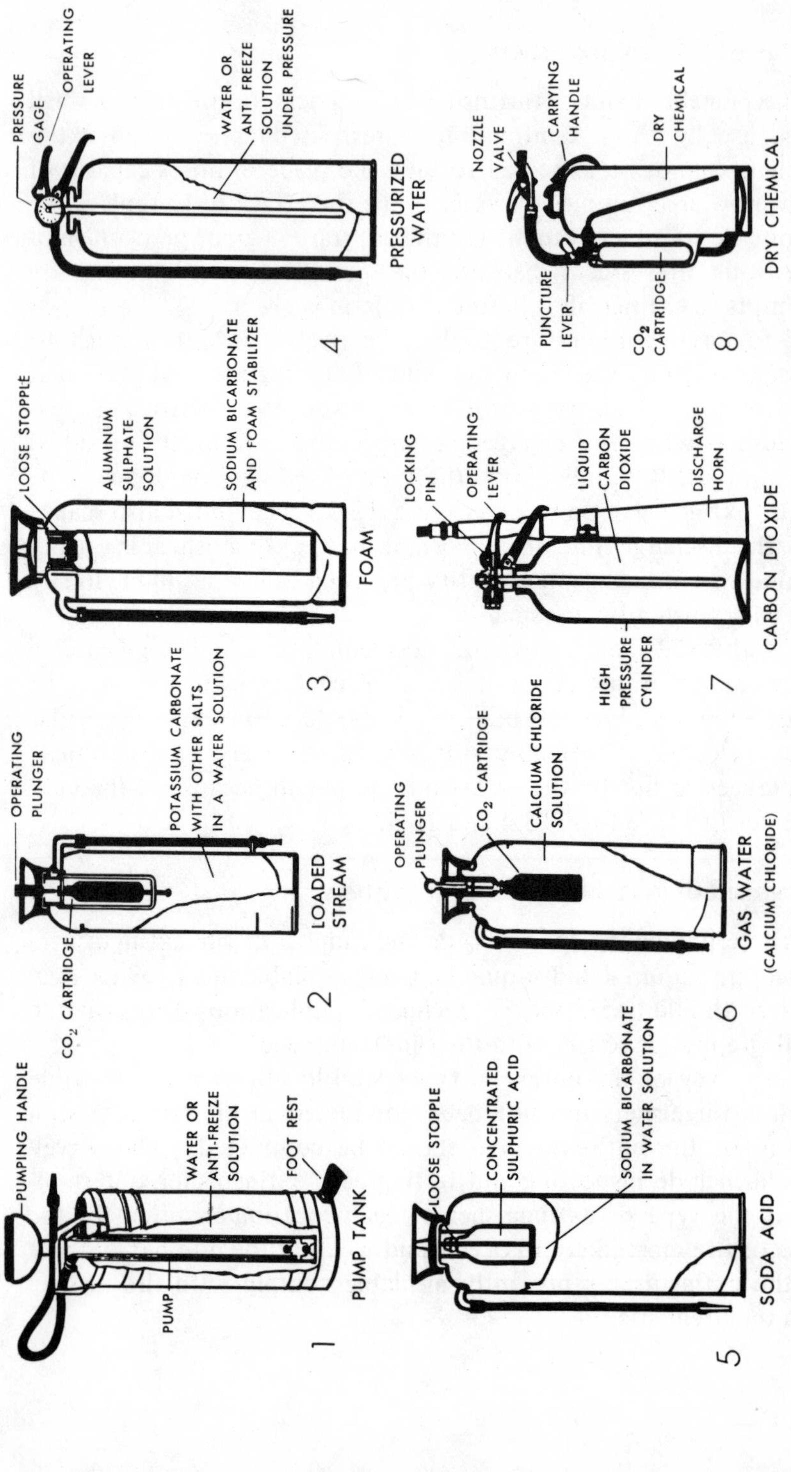

Fig. 13-3. Cutaways of the eight basic hand fire-extinguishers. Not that numbers (1) through (6) use water as the basic extinguishant. Carbon dioxide (7), and dry chemical (8), complete the family of hand extinguishers.

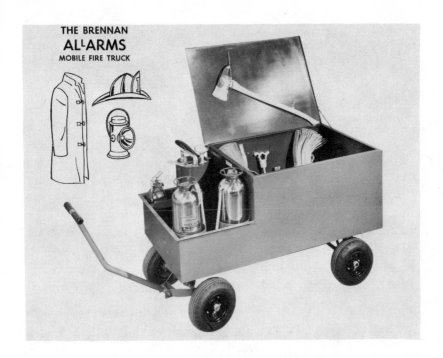

Courtesy of the Brennan Co.

Fig. 13-4. Hand-drawn "hose houses." This type of mobile fire truck can be utilized effectively within the industrial building. Firefighting organizations will assign manpower to operate them.

Fire extinguishers designed to attack different classes of fire should be placed in separate locations. This procedure will effectively reduce the possibility of the wrong equipment being inadvertently used. As an example, a water-filled extinguisher and a carbon-dioxide extinguisher should not be placed side-by-side in an area where an electrical fire is most likely to occur. A person in a moment of excitement may inadvertently, or perhaps because he is untrained, employ the water-filled extinguisher instead of the carbon-dioxide extinguisher, causing serious injury or death.

It is possible that more than one type of extinguisher is needed in a particular area; if this is the case, they should be separated,

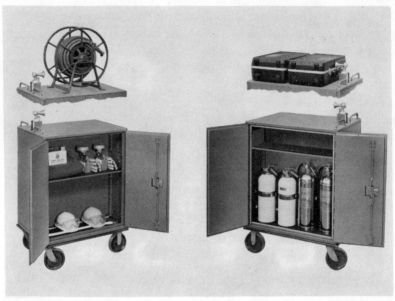

Courtesy of the Young Fire Equipment Co.

Fig. 13-5. The Infighter—front and rear views of another version of the in-plant mobile fire truck for carrying equipment to the emergency area. The type and amount of fire-fighting equipment is determined during the fire protection survey. (Top Left) A live hose reel; (Top Right) Self-contained breathing apparatus—both are optional equipment.

with each type placed as near as possible to the location where the potential of the particular class of fire is the greatest.

Marking Fire-Fighting Equipment Locations

Again, reference is made to the listing of technical data (National Fire Protection Association, National Fire Codes. See Bibliography.) which specifies the minimum marking requirements of all locations of the various types of fire-fighting equipment which is required in given areas.

The reference data specify minimum visible marking; experience has shown that additional location markings are desirable and will increase the rapidity with which the equipment can be brought into play. Fire extinguishers are often mounted on columns or uprights throughout the industrial complex and are visible only from a small percentage of the area where the extinguisher is intended to be used. If the column or upright were marked on all sides, the location of the extinguisher would be visible 360 degrees from that point.

Where extinguishers are mounted on the walls, columns, or uprights, or in areas where machinery or material obstructs the view from an eye-level line of sight, the locations of the extinguisher or other equipment may additionally be marked at a greater height than the material or machinery throughout the area. For example, standpipe hose or hand-extinguisher locations in a warehouse could more easily be identified if the wall or the column where the equipment is located were properly marked near the ceiling.

Usually, the locations of fire-fighting equipment throughout the facility are marked by red decals in the form of arrows directed toward the equipment, solid red stripes, or red and white stripes. An acceptable method for marking hand-extinguisher locations, particularly when the markings appear at a greater height than that prescribed as a minimum standard, is to use the color-coded symbols assigned to the extinguishers for use on the four classes of fire. As an example, for an extinguisher to be used on ordinary combustible materials, the location could be marked with a green triangle (Class A). In areas where live electrical equipment fires are more likely to occur, the locations could be marked in a blue circle (Class C fires). Marking of fire extinguisher locations could be carried even further by merely designating the locations with the color-coded symbols assigned to that particular class of extinguisher.

Hand fire-extinguisher locations must most often be established along the thoroughfares within production and warehousing areas, and are thus exposed to being struck and damaged or destroyed, particularly in warehousing operations where fork lift trucks are utilized. If the extinguishers or standpipe hoses are overexposed to this type of hazard, consideration may be given to protecting the equipment by erecting a substantial barrier around it. Strap iron can be effectively used to protect hand fire extinguishers without hindering their accessibility in an emergency.

Engineered Extinguisher Systems

In some facilities, the fire potential will be greater in certain areas than others because of the inherent potential of fire in given types of manufacturing processes. To combat this hazard specially engineered fire protection systems are installed in the production equipment or machinery as required. As an example, carbon dioxide extinguisher equipment is engineered for and built into large printing machines.

The facility's security director and the security force must be trained to the extent that they have a working knowledge of how these systems operate and how they can be shut down in the event of accidental activation. If their responsibilities are to include visual inspection of the specially engineered systems, their knowledge must be extended to cover the location of all extinguishant discharge heads, the lines between the discharge heads and the extinguishant supply. If gauges record the level of supply of the extinguishant, they must be able to read the gauges accordingly for unfavorable conditions.

Automatic Sprinkler Systems

The first automatic sprinkler system was devised and installed in England in the late 1800s. The system was simple but effective. It consisted merely of a series of pipes with holes, which were installed throughout the area being protected. This was a form of the dry sprinkler system. The water was held from entering the system by a valve secured in its closed position by a rope arrangement installed along the ceilings throughout the protected area. When a fire occurred, the rope was burned through and the valve was opened by the combination of water pressure and the action of a strong spring. Water entering the array of pipes was distributed over the protected area through the holes in the piping.

Automatic sprinkler systems were first installed in the United States in the early 1900s and were a little more sophisticated. However, the design of this original automatic sprinkler system has not changed appreciably. About the same components appear in the automatic sprinkler system installed today as were used in the earlier versions.

There have been improvements on valves, gauges, and some developments in fire protection alarm devices; however, the original automatic sprinkler system operated efficiently and required few changes.

There are three types of automatic sprinkler systems, and they are designed for three specific purposes. However, it is possible that all three types or a combination of any of the three may be found in a single industrial facility. The three types of systems in general use today are wet pipe, dry pipe, and deluge.

The *wet pipe system* is designed to operate by having water available throughout the entire system of pipes which is discharged immediately when a sprinkler head is ruptured.

The *dry pipe system*, uses compressed air and has valves at the riser location to prevent the water from entering the piping installed in the protected area until a sprinkler head has been ruptured, when the compressed air escapes. As the air pressure decreases, valves open to permit water to be forced through the system.

The dry pipe system is installed throughout areas where freezing may occur and damage the sprinkler system equipment, rupturing pipes and rendering valves inoperable.

The *deluge system* is somewhat similar to the dry pipe system, since no water is available at the sprinkler head until fire starts. The sprinkler heads are of the open variety and may be likened to small individual nozzles. These sprinkler heads are designed to direct the flow of water in a specific direction much the same as a deflector on a conventional sprinkler head used in the wet and dry systems.

Water in the deluge system is held at the riser by valves which are either automatically or manually operated. If the area protected by a deluge system is further protected by heat sensors or other fire protection devices, valves automatically open, and a continuous flow of water emerges from the open sprinkler heads when these devices are activated.

Whichever system or systems are installed at the facility, the security force must have knowledge of its operation and be trained in its use. Specific instructions to the security force must be formulated to include reading air and water pressure gauges at the riser locations, indicating the specific pressures that must be maintained, and the right action to be taken in the event the pressure falls below a specified gauge reading.

Sprinkler System Components

If security guards are to be properly trained to intelligently and quickly respond to various conditions which may arise with any of the types of automatic sprinkler systems, they should have a working knowledge of the basic component parts. They should also be knowledgeable about the piping that delivers the water to the fire area when the system is activated.

The schematic drawing (Fig. 13-6) shows the principal components in the automatic sprinkler system, from the water sources to the sprinkler head and, more importantly, defines the location of the various components in relation to one another. Often security guards have full knowledge of the location of yard hydrants, post indicator valves, and siamese connections, but have no idea how the components are related to one another.

A brief description of all components, which are shown in the sketch in Fig. 13-7, and their function in the automatic sprinkler system appear below.

1. *City hydrant.* Used only by the public fire department. Usually it is tied directly into handheld hose lines or tied into the pumper truck to insure maximum water pressure is maintained in the sprinkler system.

2. *Siamese connections.* Hose lines from the pumper truck are tied into the sprinkler system through a connection known as the siamese connection—often referred to as "fire department connections." This component is connected directly to the sprinkler riser and water is supplied to the sprinkler system under pressure from the pumper truck. This insures that an adequate supply of water is available to keep all ruptured sprinkler heads operating at their full rated capacity. Note that the pumper is using the city main as a source of water supply and not the private water supply lines.

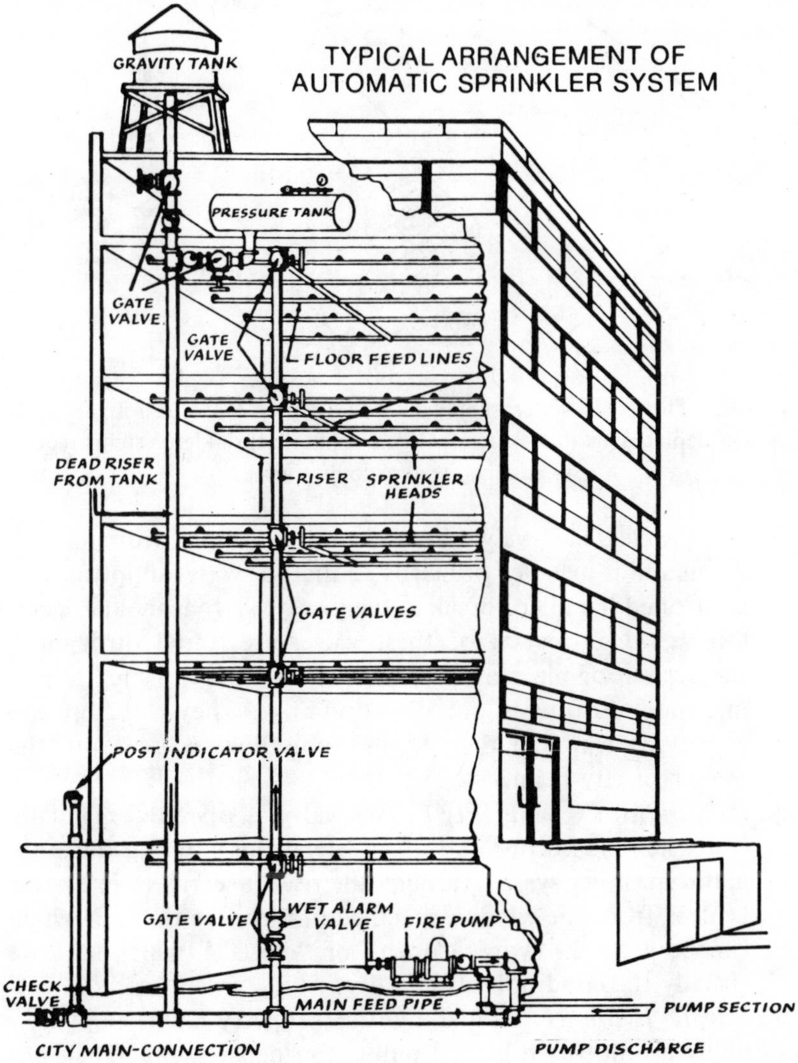

TYPICAL ARRANGEMENT OF
AUTOMATIC SPRINKLER SYSTEM

GRAVITY TANK

PRESSURE TANK

GATE VALVE

GATE VALVE

FLOOR FEED LINES

DEAD RISER FROM TANK

RISER   SPRINKLER HEADS

GATE VALVES

POST INDICATOR VALVE

GATE VALVE

WET ALARM VALVE   FIRE PUMP

CHECK VALVE

MAIN FEED PIPE

PUMP SECTION

CITY MAIN-CONNECTION   PUMP DISCHARGE

Fig. 13-6. The automatic sprinkler system is not as complicated as it first appears. Its effectiveness, however, must not be underestimated.

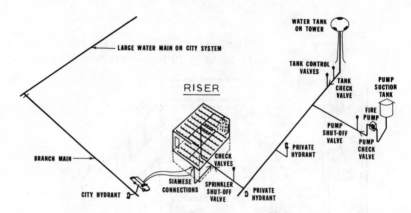

Fig. 13-7. The basic water supply lines and valves that control the water flow are depicted in this sketch. Most automatic sprinkler systems require an auxiliary water supply source.

3. *City valve.* This valve controls water supply from branch mains and is installed primarily so that the water supply can be controlled if a major break in the private system should occur.

4. *Check valve.* Several of these valves are found throughout the system of pipes and are constructed and designed to permit water to flow in one direction only. They will close and check the water in the system from flowing back into the branch or city main.

5. *Post indicator valve (PIV).* This valve is operated manually and when closed prevents any water from entering the automatic sprinkler system through the riser it services. The name evolves from the window installed in the valve through which can be seen the words "open" or "closed," which indicate visually the condition of the valve.

6. *Private hydrants.* Owned by the property owner and normally are not used by the public fire department. Their primary use is for tying hose available on site to manually attack the fire. The private hydrants are normally used by the facility's fire fighting brigades.

7. *Sprinkler shutoff.* This valve is installed in the riser and is usually in the form of an OS&Y or outside screw and yoke valve. The control wheel may be located inside or protrude

through the peripheral walls and be located outside the protected building. When the large screw is exposed, the valve is opened; when the screw cannot be seen on the outside, or only a small portion can be seen, it would indicate the valve is closed.

8. *Pump shut-off valve.* This valve is used to shut off the flow of water to the system from the auxiliary water supply source when this source is no longer needed.

9. *Pump check valve.* These check valves are similar to those described above. They check the flow of water in the system from returning or emptying into the auxiliary water-supply source.

10. *Auxiliary pump.* Almost all automatic sprinkler systems require an auxiliary fire pump to maintain constant water pressure at a level to insure that all sprinkler heads in the system, if operating, continue at their full rated capacity.

11. *Suction tank pump.* The pump's function is simply to pump water from the auxiliary source into the system in sufficient quantity to insure efficient operation of the sprinkler system. All pumps in the system are designed to start automatically and to be tested manually as well.

12. *Tank check valve.* The function of this check valve is exactly the same as the valve described in item 9, above. It checks the return flow of water.

13. *Tank control valves.* These valves control the flow of water from the gravity tank and are usually of the post indicator valve type.

14. *Water tank or tower.* The water or gravity tank contains the auxiliary supply of water for the system. This type auxiliary supply vessel is most often found in the industrial complex.

Types of Sprinkler Heads

The sprinkler head is often referred to as the workhorse of the automatic sprinkler system, because it is the sprinkler head which ruptures and discharges a predetermined amount of water directly upon the fire when a fire starts.

There are several different types and shapes of sprinkler heads and each is specially designed to be installed in particular types of places. Those appearing in Fig. 13-8 are the most commonly used in the industrial complex. The type of head and the rating of the fusible material will be determined by local requirements in each area where the automatic sprinkler system is installed.

The four main parts of the conventional sprinkler head and their functions are as follows, and are also designated in Fig. 13-8.

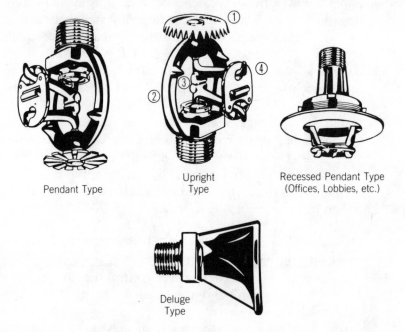

Pendant Type

Upright
Type

Recessed Pendant Type
(Offices, Lobbies, etc.)

Deluge
Type

Fig. 13-8. Four types of sprinkler heads. The basic sprinkler head has four components: (1) the deflector; (2) the frame; (3) the struts; and (4) the fusible link.

The *deflector* is designed in several different shapes. The shape will depend upon how the water being discharged is to be directed. The purpose of the deflector is to direct the flow of water striking it in a predetermined direction and configuration.

The purpose of the *frame* is to hold the deflector and the struts in position.

The *struts* consist of two pieces of metal which act as stoppers. They are positioned over the hole in the sprinkler head in such a manner that the hole from which the water will be forced is plugged. The struts are held in place rigidly by the fusible link.

The *fusible link* is a soft metal seal which melts at a predetermined temperature. The normal rating of fusible links used in an office area, for example, is 135 degrees Fahrenheit. This means when the temperature at the sprinkler head reaches this intensity the link will melt or "fuse" and pressure of the water against the unsecured struts will force the struts out of position, and a free flow of water under pressure will occur. The fusible links are manufactured at many different ratings, or melting-point levels, depending upon the specific intended application.

The majority of sprinkler heads installed in an industrial complex are generally of the type illustrated in this chapter. A brief description of each type and its function follows.

Pendant type heads get their name because they are installed in the sprinkler pipe in an inverted position; that is, the deflector is pointed downward. This type will usually be found installed under stairs, through false ceilings or similar installations. The deflector is shaped so that the water is forced upward against the protected area rather than downward.

Upright type heads are so named because they are installed on top of the sprinkler pipe with the deflector uppermost. In this case, the deflector is shaped so that water striking it is forced downward over the protected area.

A *recessed pendant head* is installed so that the entire head is recessed in the ceiling with only a part of the struts and the fusible link exposed. This type of installation is normally found in offices, lobbies, and in other areas where the decor of the surroundings is considered important. It does not appear to have a deflector when viewed from the area being protected. The deflector drops down as soon as the struts are separated by the force of the water.

The *deluge head* is entirely different from those discussed above. It consists of a deflector only. These heads are often referred to as "window" or "cornice" sprinkler heads because they are used to protect windows and other openings, including entire sides of buildings, from exposure to a nearby fire. Since the deluge head is open at all times, water entering the system will immediately be forced against the deflector head and directed in a predetermined direction.

Accidental ruptures of sprinkler heads occur very infrequently and the water damage attributed to such ruptures is infinitesimal. Ruptures are usually caused by a mechanical injury, excessive corrosion, overheating, or freezing. All of these conditions can be overcome, if proper protection to the sprinkler head is applied and the proper type of head is installed.

Standpipe Hose Systems

The standpipe hose system (see Fig. 13-9) is usually a part of the fire protection system installed in an industrial facility whether or not an automatic sprinkler system is utilized. Standpipes are installed separately from the automatic sprinkler system and the hoses in the system are not fed water through the sprinkler risers, because the water available for distribution to the sprinkler head is reserved for that purpose only. If a number of sprinkler heads are operating and at the same time hoses were operating and fed by the same water supply, one or both of the systems might be robbed of the required water because of low pressure due to the excessive drain of water.

When the standpipe system is installed in a multistory building, the vertical water pipe is normally found in the fire wells or fire escape stairs. The hoses in this case will be located on each floor level, usually just inside the enclosed fire stairs at each emergency exit door leading to the stairwell. This location is chosen so that when a fire occurs on a particular floor, the individuals operating the hoses fight the fire from a point of safe retreat, in the event the fire becomes great enough to cause them to abandon their efforts.

Hose Cabinets and Hose Handling

Hoses which are a component part of the standpipe hose system will be found stored in hose cabinets, on reels, or folded and stored on horizontal supports. Hoses stored in hose cabinets are

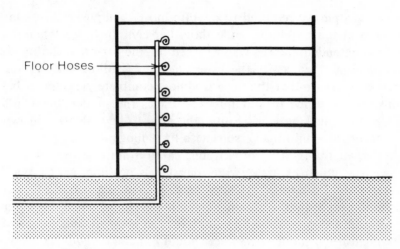

Floor Hoses

Fig. 13-9. The standpipe hose system supplements the automatic sprinkler system and draws its water supply from separate lines.

adequately protected against atmospheric conditions. However, exposed hoses throughout the production and warehousing areas are often unnecessarily exposed to caustic atmospheres and other deteriorating conditions. This type of exposure causes the material to deteriorate prematurely.

Exposed hoses can be covered with plastic or similar material, prolonging the life of the hose immeasurably. Many such covers can be purchased which are tailored to fit hoses stored on reels, on horizontal brackets, and other holding arrangements. The covers are usually a fluorescent red and lettered appropriately.

Hoses should always be connected to the standpipe so that they are immediately available for use. In some installations additional couplings may be available at the standpipe hose location to permit more hoses to be connected to the system when required.

All standpipe hoses will have a play pipe or nozzle installed on their working end. The type will depend upon the type of fire hazard likely to be encountered in the area the hose services. The nozzles may discharge a solid stream of water, or they may be of the water-fog type. Nozzles are manufactured today which are capable of discharging either a solid stream, spray, or fog merely by making an adjustment of the nozzle.

Two people are required to operate most standpipe hoses. One remains at the standpipe while the other lays the hose. When the hose is extended and the water turned on, the person at the standpipe should then assist the hose-handler. Caution must be exercised to insure that the hose is almost completely extended before the full force of water is released as the sudden burst in a folded or rolled hose will cause the hose to be violently thrown about, possibly injuring persons with its flailing nozzle.

As in all fire-fighting procedures, caution must be exercised to prevent as much excessive water damage as possible, yet as much water must be applied as is necessary to extinguish the blaze. Common sense by the hose-handler will determine when the water can be turned off. Once the fire has been brought under control (extinguished), the hose-handler should remain at the scene ready to apply water in the event the fire should rekindle. The second man of the team should remain on post at the standby valve location ready to turn the water on if rekindling does occur.

Hose Houses and Equipment

A hose house (see Fig. 13-10) should be provided for each outdoor private or yard hydrant and should be arranged so that the hydrant is as close to the front or open side of the house as possible.

The hose house should provide shelves or racks for the hose, and accommodate the hydrant and equipment. The hose house illustrated in this chapter is a typical wood frame design. There are several metal houses of different sizes and shapes manufactured. The construction should be designed to insure that when the house doors are open, all equipment is exposed and can be easily reached. Caution should be exercised to insure no equipment is on the hose shelves or racks that would be knocked off when the hoses are removed.

The National Fire Protection Association (NFPA) lists the following equipment as the minimum amount required in each industrial-site hose house:

2 underwriters' play pipes   1 extra hydrant wrench (in addition to wrench on hydrant)

1 pair play-pipe brackets

1 fire axe	4 coupling spanners
1 pair fire-axe brackets	2 hose and ladder straps
1 crowbar	1 underwriters' play-pipe holder
1 pair crowbar brackets	2 2½-inch hose washers (spares)

This list can be modified, depending upon the local situations. Almost every hose house now contains a fog nozzle in addition to the required equipment.

Hose houses should be numbered and the number placed on each piece of equipment assigned to the house to insure that equipment removed for use is returned to the right place.

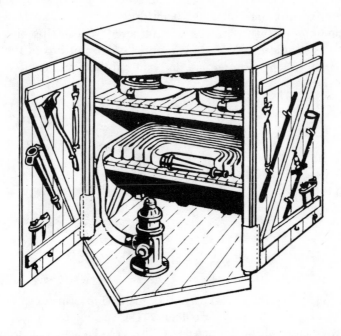

Fig. 13-10. The hose house, whenever possible, should be so located as to permit the hose to be coupled to the yard hydrant. Inspections must include an inventory of equipment.

In some instances, hose houses cannot be placed over the hydrants because the exterior of the property protected has been landscaped and the houses are considered unsightly. In such instances, the houses should be located nearby and should be maintained and inspected just as though they were right at the hydrant.

Too often hose house housekeeping is very poor. Dust and dirt collect, the equipment is not maintained properly and soon deteriorates, weeds grow through the floors and become entangled in the equipment, metal houses are allowed to rust and houses are often damaged by being struck by vehicles.

Hose houses should be closed and sealed or secured with a "breakaway" padlock. This type of padlock has a serrated shackle which can be easily broken. Railroad seals or wire and lead seals may also be used (see Fig. 13-11). The seals or padlock will indicate whether or not the house has been opened without authority and equipment possibly removed or tampered with. Hose house inspections should be covered by a report on conditions of the house and state of the equipment and should be conducted at least monthly.

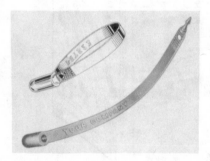

Courtesy of E. J. Brooks Co.

Fig. 13-11. Inexpensive seals such as these are used to secure electrical boxes, hose houses, or emergency exit doors. The guard force inspections will quickly establish whether or not the seal has been broken or tampered with and the object entered.

An inspection must, of course, insure that all equipment is in its place. There are two commonly used methods to help the security guard inspecting the hose house account for all the required equipment. One method is merely typing a list of the required equipment on a 3 × 5 card, laminating the card with a laminator used to produce identification cards, and making it available in the hose house. Another method is to paint a silhouette of the required items on the shelves or doors in the location where the item should be. Items will be conspicuous by their absence.

Fire Doors

Fire doors are provided to help arrest the spread of fire between departments or areas of a facility, and they are made and installed to certain standard specifications. Doors for fire protection will be metal clad. These doors will be held in their open position either by a fusible link or by an electromagnetic device which may be tied into another component of the fire protection system, such as a water-flow alarm at an automatic sprinkler riser, a heat sensor, or a similar type of device.

Fire door maintenance and inspection are essential and should include at least the items appearing below:

1. Doors must be operable at all times. A continual inspection and maintenance program is necessary to insure that doors can be closed or will close and latch when fire occurs.
2. It is necessary to keep surroundings, including door openings, clear of everything which would be likely to obstruct or interfere with operation.
3. When necessary, a pipe or slat framework should be provided to prevent any piling of material against sliding doors which would impede operation.
4. Doors must be kept closed and latched or arranged for automatic closing. Blocking or wedging open a door makes the door inoperative.
5. Automatic or self-closing devices must be kept in proper working condition at all times.
6. Hardware should be examined frequently and any parts found to be inoperative promptly replaced.
7. Hinges, catches, latches, and stay rolls are especially subject to wear and if found defective should be repaired or replaced.
8. Guides and bearings should be kept well lubricated to facilitate operation.

9. Any breaks or tears in the sheet metal covering should be promptly repaired. Metal-clad doors should be inspected for dry rot.

10. Fusible links or other heat-actuated devices should not be painted. Care must be taken to prevent paint accumulation on stay rolls.

11. All doors normally held in the open position should be operated at frequent intervals to insure proper operation.

12. Cables or chains employed on suspended doors should be frequently inspected for excessive wear and stretching. Chains or cables of bipartite counterbalanced doors need frequent adjustment to insure proper latching and to keep the doors in proper relation to the door openings.

13. Doors should be closed at night, Sundays, and other times when the openings are not in use, if not equipped with automatic closing devices. An automatically operated door holder is shown in Fig. 13-12.

14. Windows of stained glass should be replaced with wired glass not less than ¼-inch thick, well embedded in putty.

Fire Protection Alarm System

The primary reason for installing an effective fire alarm system in a facility is to save lives, and the secondary purpose is the protection of property. A properly installed integrated fire alarm system will accomplish both with efficiency. The object of the modern local fire alarm system can be described in two words— "early warning." This includes early warning of fires occurring in hidden or unoccupied locations, early warning to all personnel in the affected area, early warning of the fire's exact location, and early warning to the firefighting service. Figure 13-13 shows a photoelectric smoke detector and Fig. 13-14, an infrared flame detector.

Thus, early warning is to insure the immediate evacuation of all personnel from the affected area and the immediate response by firefighting crews which may mean the difference between a minor fire and a major conflagration.

Courtesy of Pyrotronics, Inc.

Fig. 13-12. The smoke and fire door holder is tied electronically into the fire protection system. The door is closed automatically when an alarm in the area is activated.

The fire alarm system components fall into three major categories:

1. Fire detectors
2. Supervisory equipment or water-flow detector switches
3. Signaling devices.

Fire Detectors

There are several types of fire detectors.

The *ionization detector* is a relatively new type of detector which operates by means of a small current passed through the air between two plates (see Fig. 13-15). Hydrocarbons and other

Courtesy of Pyrotronics, Inc.

Fig. 13-13. The photoelectric smoke detector is used in hallways, stairwells, air ducts, and areas where smoke is likely to be driven during the second (smoke) stage of a fire.

invisible products of combustion emitted by a fire during the first, or incipient, stage interfere with this current and cause an alarm.

This detector is extremely effective because it is capable of giving a very early warning of fire. It is particularly suited to protection of computer rooms, "sterile" rooms, and areas where spontaneous combustion may occur, as well as in all types of air transfer ducts. The ionization detector will adjust itself to normal amounts of tobacco smoke, but will still react to the first signs of fire.

The *thermal detector* senses temperature. Some thermal detectors work at a fixed temperature rating and others, referred to as "rate-of-rise" detectors, react to an unusually rapid rise in

Courtesy of Pyrotronics, Inc.

Fig. 13-14. The infrared flame detector is used in the overall fire protection
system. Note that a pilot light is also included.

temperature. Some detectors may combine both of these characteristics; perhaps the combination type is the most sensitive
(see Fig. 13-16).

The sensitivity of the combination fixed temperature and rate-
of-rise detector permits its installation over a widely spaced area,
resulting in a saving on labor costs for installation. The fixed
temperature detector, somewhat less expensive per unit, must be
more closely spaced. Fixed temperature detectors are normally
used in closed areas, such as mail rooms and record storage areas,
where there is no need for a more sensitive detector. They are
also effectively used where rapid temperature fluctuations make
the rate-of-rise detector impractical for installation.

Courtesy of Pyrotronics, Inc.

Fig. 13-15. This device, known as the "Ionization Fire Detector," is activated when a fire is in its first (incipient) stage.

Pneumatic detectors are similar to the rate-of-rise detectors. This operation includes an airtight tube which is installed throughout the protected area. A rapid increase in temperature raises the pressure in the tube and triggers a pneumatic switch associated with the tube which starts the alarm. The pneumatic detector has the advantage of safety from explosions and is suitable for high-ceiling rooms and other areas where access is difficult. The pneumatic switch or triggering device can be installed out of the explosive area or in an area easily accessible for inspection and service.

Supervisory Alarm Devices

Sprinkler water-flow detectors are the second category of components in the integrated fire alarm system. When installed in the

Courtesy of Pyrotronics, Inc.

Fig. 13-16. This fixed temperature and rate-of-rise thermal detector contains a signal light to indicate that a particular detector has been activated.

automatic sprinkler system, water-flow and pressure switches activate an alarm when the system goes into operation. The water-flow detectors will normally be found in the main risers and consist of a flexible paddle which is driven in the direction of water flow when it occurs, switching an electrical contact which signals the alarm.

The automatic sprinkler system can also be protected by the following supervisory devices:

1. *Temperature supervision* to warn when the auxiliary water supply tank or pipes are near freezing temperature.
2. *Air pressure supervision* to activate an alarm when air pressure drops in pressure tanks or dry pipe sprinkler systems.
3. *Tank water level supervision* to warn when the level in a water tank falls below the minimum established.
4. *Post indicator valve supervision* to warn when the main water supply valve to a sprinkler riser is shut off.

5. *Gate valve supervision* to warn when a secondary water supply valve is shut off. This type of supervision should be installed on the OS&Y valve at the automatic sprinkler riser.

Signaling Devices

The signaling devices are the third major category of the system's components. The basic signaling arrangement may operate in this fashion: A detector or supervisory device transmits a signal. A zone transmitter for that specific area receives the signal and transmits it to a central panel. The central panel "reads" the signal and immediately activates the appropriate audible and/or visible alarms.

Coding alarms means providing an automatic means of announcing information by signal whether it is an alarm, a trouble warning, or an all-clear message, and whether or not it comes from the fire alarm system or the sprinkler supervisory system. The specific zone which is affected is, of course, also identified.

Coding can be done by means of a patterned series of bell strokes or by lights on a panel or special buzzers, horns, or sirens. The fire protection equipment may include apparatus to produce printed records of all signals delivered (see Fig. 13-17).

Manual Fire Alarm Stations

Manual fire alarm stations are manufactured in numerous different designs. The stations are wall mounted and will usually be classified as pull boxes; that is, a lever or trigger held in place behind breakable glass; a seal, or a small rod must be manipulated in order to initiate an alarm signal. These will normally be of the "single action" type. This means that a single pull of the operating lever will initiate the alarm. The manual fire alarm stations are classified into two categories.

1. *Local alarm stations* are overprinted in bold letters with the word "local." This type of station activates an audio alarm locally only. It does not transmit a signal to the location where the fire alarm protective system is monitored. The purpose of these stations is to alert and/or evacuate personnel

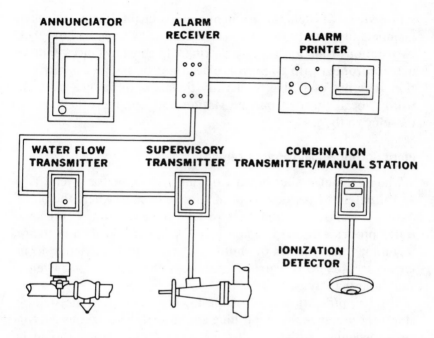

ANNUNCIATOR   ALARM
                  RECEIVER
                                      ALARM
                                     PRINTER

WATER FLOW   SUPERVISORY   COMBINATION
TRANSMITTER   TRANSMITTER TRANSMITTER/MANUAL STATION

IONIZATION
DETECTOR

Courtesy of Honeywell, Inc.

Fig. 13-17. A basic fire protection alarm system will usually include the alarms and components depicted here. More sophisticated systems may utilize fire door closers, deluge system valve activators, and many other accessories.

during drills or for other reasons, without altering the fire-fighting organization.

2. *Remote signaling stations* do not contain the word local and, when activated, transmit a signal directly to the monitoring panel. It is possible that this type of station could be activated without sounding an audio signal locally. Normally, however, an audio signal will usually occur simultaneously with the signal being transmitted to the monitor.

These signaling devices can be, and usually are, wired to the coding arrangements. When either type of alarm station is activated, the coded signal assigned to the zone protected by the particular alarm will be automatically activated.

Interviews with employees in numerous facilities and with many members of security forces revealed that knowledge of how these two stations function is sadly lacking. In most instances, individuals confronted with two boxes placed side-by-side in a demonstration were not even able to identify the major difference, since the word "local" printed on one station and not the other had no meaning to them.

Plant Shutdown and Housekeeping

When a plant is shut down from all operations for several days or for extended periods, as occurs in many cases where the procedure is to allow all employees to take their vacation at the same time, proper safeguards against fire assumes added importance.

During these periods of shutdown, it is usually advisable to increase security force patrols throughout the facility on a twenty-four-hour-per-day basis.

The fire protection plan of the facility should include step-by-step procedures in accomplishing shutdown. This will usually involve assigning duties to all department heads which should be supervised by the facility's fire brigade chief. The plan must be monitored throughout the shutdown period by the security force.

To attempt to assemble here a checklist of items to be considered appears hardly worthwhile, because to be of any great assistance, the length of the list would be very long and then probably would not cover all areas or items. However, such a checklist should be formulated to meet individual company needs and it should be included in the written fire protection plan for the facility. The list should be divided into areas of responsibility, and the protection plan should designate certain individuals who will be responsible for conducting a detailed inspection of all areas to insure that the shutdown procedures have been properly applied.

Similarly, it would be useless to attempt here to assemble a list of items to be checked by the security force or other responsible individuals covering good housekeeping. There is an old familiar adage, however, which sums up the extreme importance of good housekeeping practices in planning industrial fire safety: "A clean plant seldom burns."

Members of plant management touring the plant in the course of their duties or on inspection tours normally restrict their movements to the main production and warehousing areas. To them the plant may appear to be well kept and clean. However, detailed inspections will often reveal that accumulations of dirt and combustible trash in off-the-beaten-track areas exist. Obviously, when combustible trash is allowed to accumulate in hidden areas, the fire hazard is even greater, because a fire may not be readily discovered. Trash containers with metal self-closing tops (see Fig. 13-18) are well worth considering as an inexpensive, effective fire-protection device.

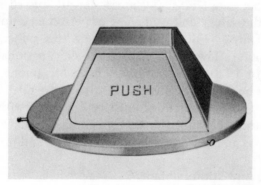

Courtesy of Sargent-Sowell, Inc.

Fig. 13-18. Metal self-closing tops that can be installed on 55-gallon metal drums are an inexpensive, effective fire-protection device.

Security force instructions should contain a checklist of items or areas to inspect in the course of their patrols through the plant. This checklist must be tailored to each individual facility.

The plant fire safety plan must also include specific instructions to those departments which are responsible for eliminating or reducing fire hazards reported by the security force. Hazards must be reduced as soon as they are reported, if the fire safety plan is to be as effective as is humanly possible.

CHAPTER 14

Safety for Personnel

In preceding chapters, many areas were discussed which relate directly to the safety and well-being of personnel employed in the industrial facility. Further discussion on this subject will be restricted to only a few areas, because the Occupational Safety and Health Act of 1970, a Federal law, is probably among the strictest laws ever to be passed by Congress. It tailors regulations to specific industries. The regulations must be complied with; it suffices to say here, time is getting short where companies have not secured the regulations and initiated safety adjustments and programs.

Parking Lots and Thoroughfares

Safety in employee parking lots and on the interplant thoroughfares can be considerably improved if speed limits are established. A sufficient number of signs must be strategically placed indicating maximum allowable speeds. The interplant thoroughfares should include traffic control signs such as stop signs, yield signs, and such other controls as are necessary. The signs erected along the thoroughfares intended for traffic control should in all cases use the wording, color, and shape of the signs that are predominantly displayed on the public thoroughfares of the state in which the industrial facility is located.

Speed retarders can be an effective tool in controlling excessive speed in employee parking lots and on blind or partially blind road intersections in the interplant network. The construction of these speed limiters is similar to those often found in shopping mall parking lots and thoroughfares.   They are constructed of cement or other roadway material formed in a hump across the roadway at a height sufficient to require drivers to reduce speed to five miles per hour or less. This construction may be painted in yellow or have caution sign installed on one or both ends to insure that drivers are aware that it exists.   Be certain the retarder does not itself become a safety hazard.

Traffic cones, similar to those used on public thoroughfares, can be purchased or, in some instances, rented. These traffic cones will materially assist directing traffic flow or restricting traffic when temporary conditions change the normal permanent pattern.

Plant entrances and exits which occur at a junction of the private road and a publicly traveled thoroughfare often create hazardous traffic conditions during shift changes. When this situation exists, security guards will normally have to be deputized by the police authority if they are to direct traffic on the public thoroughfare. The local laws should be researched to determine whether or not this requirement is necessary.

In many instances, the state, county, or city governments will install a traffic light to be operated only during heavy traffic periods, as at shift changes.   These traffic lights can be automatically operated to change signals or can be manually operated by a security guard during these periods, thereby adjusting the cross-traffic conditions.

A study and analysis of this type of traffic problem may indicate that only warning signs are necessary on each side of the public thoroughfare and plant roadway junction.   Whatever method or procedure is finally developed will depend upon the volume of traffic, the type of traffic, and the normal speed of the traffic traveling the public thoroughfare.

Security Force Instructions

Security force instructions must include guidelines or checklists in the plant safety area to be inspected daily in each individual facility. The items and areas on the checklist are usually restricted to commonplace safety hazards which a normally prudent individual would observe. In some instances, a security force with proper training could be assigned more technical safety inspections, which might include more detailed inspections of machinery guards, facial protection installed permanently on machinery, and other items of this nature.

A properly trained security force can assist immeasurably in policing and enforcing the safety plan. Consideration should be given to permanently assigning the chief or supervisor of the security force to the plant safety committee. He, probably more than any other single individual employed at the facility, will have a detailed knowledge of the conditions existing in every area of the plant, however small and remote the area may be.

CHAPTER 15

Further Aspects of Theft and Pilferage Control

Almost all of the areas previously discussed and the security measures which may be applied relate to the control of theft and pilferage. The degree of control established depends entirely upon whether or not the study was detailed and an intelligent analysis revealed the degree of security which must be applied. If this function was satisfactorily concluded, and the physical aids or controls will be incorporated into the protection plan, a great deal will have been accomplished to eliminate or substantially reduce theft and pilferage of company property.

The discussion which follows pertains to those areas not previously examined. However, they may be closely related operationally within a particular facility.

The Undercover Agent

The undercover agent can be an effective aid to the security director in many ways. Briefly, the undercover agent is an individual normally employed from or through a contract security or investigative company, private or public. His services are secured through the plant's regular employment procedures, and he is placed in the affected department in the industrial facility.

The decision of whether or not an undercover agent is needed to determine or develop specific information is usually based upon losses which have not been eliminated by implementation of the physical protection plan.

The presence of the undercover agent and his assignment must be known to as few individuals in management as possible. Normally, the security director and the plant manager are the only two individuals with this knowledge. The success of the agent's mission depends, first, upon the secrecy of his intentions and, second, on his professional ability.

The method used in actually placing the undercover man on the payroll is often the most critical phase of the investigation. In many cases it is not prudent to take the personnel department or the person who normally does the hiring into the circle of those who are aware of the investigation. In recent years, it has been relatively easy, and in many instances, preferable, for the undercover man to get himself employed without help from his management contact in the organization. The tight labor market, together with the increasing rate of employee turnover, has made this method more workable and reliable. The obvious hazards are that the undercover man may not appeal to the person doing the hiring, or once being hired, the man may be placed in the wrong area within the organization. Often it is possible for him to effect a transfer on his own after proving himself to be a good worker—in some instances his management contact can move him without involving anyone else in the operation, once the undercover man is on the payroll.

Obviously, it is sometimes necessary to include enough people in the informed inner circle to get the operative hired and placed in the proper position.

Once on the payroll the operative is expected to report his findings by means of a daily narrative-style report which is written immediately after work while events are still fresh in his memory.

The importance of reports on an undercover investigation cannot be overemphasized. Without them, the full advantages of the program cannot be realized, no chronological history of the investigator's efforts and observations would exist and those seemingly unimportant details that have a tendency to tie together loose ends would be lost for all time. A well-written report will usually contain more information than is needed.

The undercover operative will usually give most of his attention to theft problems, or the possibility of theft problems, by virtue of his training; however, he should also be alert to other problems which are becoming increasingly severe, such as sabotage, vandalism, and drug abuse within the work force.

The undercover agent can be effectively employed and is of tremendous assistance in many areas. Some of these are:

1. To find the methods being used to steal and remove company property from the protected area
2. To identify positively, by securing factual and physical evidence, those individuals involved in illegal activity
3. To conduct a morale and attitude survey of a certain group or groups of employees; this information may prove invaluable in assisting management to formulate correct decisions
4. To determine whether or not, or to what extent, employees may be using illegal drugs or consuming alcoholic beverages on company property, particularly during second- and third-shift operations
5. To find out whether or not employees are gambling; that is, whether or not certain employees are involved in "the numbers game," soliciting off-track bets, or merely selling chances on various sporting events
6. To determine who is responsible for acts of sabotage, however minor, which are being committed
7. To test the effectiveness of the security plan in effect.

The undercover agent is often employed after a physical security survey reveals that major thefts are being perpetrated. In such a case, it is usually wise to employ the undercover agent in advance of effecting any physical changes in the protection plan. If a new protection plan is formulated, or the existing plan revitalized, the thief will normally merely adapt his *modus operandi* to circumvent controls established by the change. It is far better to identify the thieves, expose their method of operation, terminate their employment or possibly prosecute them, and only then to reorganize the protection plan so as to eliminate the exposed loophole.

Trash Removal

The security force must constantly keep trash removal operations under scrutiny and surveillance. Modern methods of trash removal were discussed in Chapter 7, but emphasis must be placed on periodic inspections of these operations by the security force. In instances where company or private vehicles are used to move the trash from the protected area, security guards should be assigned on an unscheduled basis to accompany these vehicles and observe as material is being unloaded.

Custodial Service Employees

In addition to the restrictions on custodial service employees discussed in Chapter 9, the following controls should be used:

1. A specific area should be available where the outer garments must be left while they are working within the facility.
2. They should be observed by the security force on an unscheduled basis while they are working within their assigned areas and are moving to and from the point where the accumulated trash is discarded.
3. Keys issued should be restricted to those actually required, and they should be recovered before employees depart from the facility.
4. The contract company should be required to submit brief personal histories of each individual to insure that none are employed on a regular basis by a competitor in your industry.
5. If classified or proprietary operations are in progress at the facility, the individuals assigned by the contract company should not be otherwise assigned to work in the facility of a competitor.

Hand Tools

Hand tools, particularly the more expensive electrically driven kind, should be marked clearly with a code or the company's designation, in such a manner that the marking cannot be easily obliterated. It may be advisable to adopt a policy of lending this type of tool to employees for their personal use to reduce attempted thefts. In this event, obviously, a receipt and recovery procedure must be implemented.

Packages

Packages carried by employees should never be authorized beyond the employee control point. Often employees, particularly in urban areas, will make purchases in local establishments during the lunch period. If this occurs and the employee uses public transportation in moving to and from the facility and his home, it may be necessary to establish a parcel check procedure at the gatehouse. Security guards should issue claim checks for the packages to insure recovery by the rightful owner.

Scrap

Selling scrap material to employees often creates problems if the scrap is not easily distinguishable from usable material, or if it can be used as a cover to pilfer usable material. Wherever possible, some means of wrapping or boxing the material should be devised, and control passes should be issued by certain specified individuals who will be authorized to release the material.

Postal Thefts

Theft through the postal system is rare but does occur. The infrequency of thefts by using the mail can be attributed to many reasons, but the dominant one is probably the fact that individuals apprehended in this sort of theft will almost surely be prosecuted by the federal government, regardless of any lenient attitude assumed by plant management. Mailroom operations, including the vehicle and method used to transport packages between the facility and the servicing postal installation, must be closely examined and should be regularly inspected. These examinations and inspections should include the possible abuse of the postage meters.

Construction Equipment

Heavy construction equipment is often used to remove stolen material from the facility, when production and warehousing areas are in close proximity to the construction area. Inspections of this heavy equipment may be quite difficult but in some instances must be accomplished.

Take, as an example, losses which started at a company when construction commenced and continued at an unabated rate, even though inspections and additional controls were applied. The thefts were finally uncovered by an undercover agent employed in cooperation with the contractor to work on his force. He discovered that two operators of a piece of equipment commonly known as a "cherry picker" were the guilty parties. The cherry picker has a basket which is elevated to permit work at high levels. The stolen items were merely dropped into the basket, the basket elevated, and, even though the security force was inspecting the machine, none of them thought of searching the basket.

Truck Shipments

Protection of intercompany truck shipments is often overlooked in formulating the protection plan. Thefts and pilferage in large quantities often employ the trucks assigned to interplant shipments; that is, shipments made between two protected areas which involve use of public thoroughfares. The security of these shipments can easily be controlled through the proper use of seals, as discussed in Chapter 7. These transport vehicles are normally driven by company employees, which increases the possibility of driver-employee collusion at either or both the point of departure and the destination. Frequent reexamination of the established procedures should be planned to insure there is absolutely no deviation.

Theft of Time

Theft of time, in the form of employees' double-carding, is not uncommon, particularly in large facilities. It involves employees punching a friend or relative's card while he is actually absent from the facility. This form of stealing is often costly and can usually be controlled effectively by stationing a security guard in the time clock location during shift changes. Patrolling security guards should be particularly observant of time clock locations, whether they entail registering payroll information on a piecework basis or straight time, to insure that employees do not violate the timekeeping policies and procedures.

Past experience has proved that if all considerations previously discussed are properly analyzed and physical security measures are tailored to fit each specific area, theft and pilferage of company property will be effectively reduced.

If the overall protection plan is executed through the employment of a professional, well-trained contract security force, fraternization and familiarization normally present when in-house security guards are used will not exist. Policies, procedures, rules, and regulations developed by management can and will be enforced and supervised. Infractions of these rules by individuals or groups of employees can be reported to management without fear of retribution; and then management must initiate whatever disciplinary measures it feels are appropriate.

CHAPTER 16

Emergency Evacuation and Disaster Planning

To effect safe and rapid evacuation of a facility when emergencies occur, a plan to cope with disasters must be developed and put in writing. The facility's security director will always be placed on the staff developing this plan.

The plan itself, particularly as it applies to procedures which will be implemented in the event of civil disturbances or bomb threats, must be considered highly proprietary information. Therefore, the finished plan must be adequately protected.

If conditions dictate that a specific plan to deal with civil disturbances must be formulated, it would be wise to provide a separate document dealing with these conditions. This document should not be incorporated in the emergency evacuation and disaster plan, as the number and category of management personnel likely to become involved in a civil disturbance would probably be only a few individuals. The civil disturbance plan would not have as wide distribution as an emergency evacuation and disaster plan since, in all probability, few, if any, of the hourly employees in a facility would be included in implementing it.

In contrast, to be effective, appropriate parts of the plans dealing with emergency evacuation and disasters must receive wide dissemination among all employees. The emergency evacuation portion of the plan in particular must be known to every employee in the facility.

To assist in developing a plan, this chapter is laid out so that it can be used directly as a guide for outline and format when drawing up an emergency and disaster planning manual at a given company.[1] The contents are based on the results of many actual applications which have been tested and proved to be effective in practice. The format is designed for use at practically any type of industrial facility, except that department designations and the title of key personnel will, in all probability, be changed. In tailoring the plan to the local situation, other minor changes will also have to be made. However, if the format is adhered to closely and the facility plans are correctly drawn and kept up to date, a complete, workable emergency plan can easily be developed.

   It must be remembered that the emergency evacuation and disaster plan must be very detailed. To generalize will invite participants to quickly assume a lackadaisical attitude. The plan must be tested initially and adjustment made if necessary. It is imperative that the plan be updated as required to insure that it does not lose its effectiveness.

---

[1] For permission to reproduce this manual, write to: Industrial Press Inc., 200 Madison Avenue, New York, N.Y. 10016.

EMERGENCY EVACUATION
and
DISASTER PLANNING MANUAL

EMERGENCY EVACUATION
and
DISASTER PLANNING MANUAL

Table of Contents

---

[2]Appendices are listed on this contents page only for completeness to show the reader the kinds of annexes a manual contains. Examples of these particular appendices will not be found in this book.

I. Introduction

The purpose of this manual is to provide a plan for handling emergencies which may occur. Although it is impossible to foresee all kinds of emergencies, some which might occur from within the plant are fire, explosion, and structural collapse.

Additionally, there is always the danger that tornadoes, high winds, or fires in neighboring industrial plants will jeopardize the safety of life and property, and the plan must cover these possibilities. An emergency may also arise from wind blowing gas into the area as the result of an accidental discharge from a neighboring facility. Although it is highly improbable that any density of gas will remain in the area for any length of time, a section of this manual will deal specifically with this possibility.

Objective: The objective of this plan is to minimize and, so far as is possible, to prevent personal injury and property damage both in the plant and the vicinities immediately outside.

This plan describes the organization, duties and responsibilities, and methods and procedures which will be initiated, and the equipment and facilities which may be required should an emergency occur. The plans are carried out only by individuals who are qualified for the responsibilities because of their knowledge, skill, and ability, to direct others during emergencies.

Civil Disturbances: For the past few years, civil disturbances have been experienced in many parts of the country, and they could conceivably erupt again. Since preparedness to cope with this type of emergency differs somewhat from other procedures, and since this information is considered as proprietary by plant management, the plan with respect to riots or other civil disorders is not included in this manual.

II. Organization

Chain of command (reporting relationships) and responsibilities. In handling any emergency, the normal reporting relationships of the plant will be adhered to as much as possible. The plant manager will be in charge, with each of his staff department heads carrying out responsibilities most closely related to their normal operations.

In the event the plant manager is absent from the property, the production manager or equivalent will assume the responsibilities assigned to the plant manager.

The emergency organization and responsibilities of the plant manager and his staff department heads are as follows:

1. Plant Manager

   a. Activates the emergency plan if it is not already activated
   b. Coordinates and directs emergency operations of the staff
   c. Communicates with appropriate corporation officials
   d. Declares when an emergency has ceased and directs plant's functions to insure speedy resumption of normal operations.

2. XYZ Department Manager

   a. Acts on and contains the emergency within his area of responsibility
   b. Immediately shuts down appropriate operations or facilities
   c. If his department is not affected, authorizes assistance to other departments including use of emergency equipment assigned to his area
   d. Accounts for departmental personnel, to insure all employees are notified
   e. Coordinates plant or area evacuation if required
   f. If required, coordinates search to insure all personnel are out of the endangered area
   g. Renders progress or action reports to plant manager
   h. Coordinates start of operations after emergency is over and implements his departmental salvage plan.

3. Maintenance Department Manager

   a. Performs immediate repairs or provides other assistance to contain or help contain the emergency
   b. Accounts for maintenance personnel to insure they have assumed their assigned duties, if the situation permits
   c. Accounts for and allocates equipment available for use by members of his department

(2)

   d. Recommends to plant manager the need for additional assistance or equipment as needed

   e. Performs repairs and salvage to insure start of operations without undue delay after emergency is over

   f. Conducts inspection of plant and renders report to plant manager on extent of damage. Estimates by departments, when operations will recommence

   g. Recommends to plant manager priority of repairs based on extent of damage to various departments.

4. Warehouse, Shipping, and Receiving Department Manager. Besides assisting generally in the emergency the department manager has the following responsibilities:

   a. Arranges for additional transportation which may be required to remove undamaged products

   b. Requests such additional security personnel as may be required to prevent thefts or looting

   c. Arranges for and organizes additional assistance he receives to effect speedy salvage and minimize loss.

5. Industrial Relations Department Manager

   a. Communicates the nature of the emergency to key personnel who are at home

   b. Administers first aid as required

   c. Arranges for outside assistance as directed by the plant manager

   d. Communicates with appropriate corporate personnel as liaison for plant manager

   e. Communicates with news media and insurance organizations

   f. Arranges for additional security guards, as required, by communicating with the contract security service organization.

6. Inventory Planning Department Manager

   a. Coordinates with warehouse, shipping, and receiving department to insure accurate inventories are being maintained

   b. Assists in directing salvage operations of the finished product to insure minimal losses

   c. In coordination with warehousing department, appraises plant manager of condition, amount, and disposition of stock.

7. Mechanical Department Manager

   a. Organizes full resources of department and commences immediate repair as directed by the plant manager

   b. Communicates with outside suppliers to obtain necessary replacement parts

   c. Communicates with utility companies to restore service as soon as practical after the emergency

   d. Coordinates activities with the maintenance department to insure that efforts of the two departments do not overlap

   e. Causes inspection to be conducted and renders report to plant manager of extent of damage and estimate of length of time to restore operations by each department involved in the emergency.

8. Accounting Department Manager

   a. Assists in accounting for plant personnel in plant during and after an emergency

   b. Assists in communications with local firms and agencies

   c. Assists in first aid and plant evacuation

   d. Communicates with plant employees and assists in reassignment of duties when required

   e. Arranges for meals for on-site personnel if required.

9. Industrial Engineering Department Manager

   a. Assists plant manager in assessing damage and in restoring normal operations

   b. Conducts simulated emergencies to insure personnel are prepared to act in real situations

   c. Reviews and analyzes actual emergencies which may occur or which have occurred in like installations to facilitate recovery operations

   d. Recommends and initiates action to revise operating procedures or modify certain facilities to minimize risks

e. Insures by daily observation that emergency evacuation routes and exits are not blocked on permanent or semi-permanent basis.

10. Alternate Management

If the plant manager or any department manager is not available during an emergency, the following alternates will take the positions of leadership to insure that the responsibilities are carried out. (Note: the list should be continued, when developing an actual manual, to designate substitute leaders for each department)

a. *Plant Manager*
Alternates:
1. Production Manager
2.

b. *Maintenance Department Manager*
Alternates:
1.
2.

11. Emergency Headquarters

Any headquarters locations may be established during an emergency, depending upon the extent and possible effects of the emergency.

a. If the emergency is confined to a small area within the plant and the assigned offices are tenable, no movement from the plant need be made.

b. Building "X"—This building will be used, if necessary, to house emergency command stations, and the plant manager will be located as follows: Location _____
_____ Phone Number _____
The location of the command post will, in most emergencies, be obvious. Notification of where the command post location is will be made verbally through the security force, and it is the plant manager's responsibility to notify the security force director of his decision. Any location

other than the above, which may be used as an emergency command post, will be announced in a similar manner.

The plant manager should detail at least three individual employees to act as his runners as soon as he is aware an emergency exists. These individuals must be kept informed of his location at all times.

III. Emergency Team

A. Members. An emergency can occur at any time and at any place in the plant. With this fact firmly established, the company emergency team is composed of suitable personnel to cover the separate work shifts throughout the working week. These are the people who will be relied upon to take immediate and effective emergency action and upon whom will be focused emergency orientation and training. They have been selected from among all employees because of the length of time they have been employed in their individual departments and their known stability during emergency situations. These people will be relied upon to perform such emergency duties as tripping off electrical current, and shutting off fuel and gas lines, etc. The positions and names of the emergency team members are as follows:

Title	First Shift	Second Shift
Chief building engineer		
Chief of security force		
Technician—maintenance department		
Technician—mechanical department		
(Continue through the number required.)		

B. Leadership. The emergency team is led by the chief building engineer, who is under the direction of the plant manager.

1. The chief building engineer will act as overall coordinator during an emergency. A backup coordinator will be his second in command. The reasons the chief building engineer has this assignment are:

    a. He has an overall knowledge of all the equipment and facilities within the plant.
    b. An independent outside telephone is located in his office.
    c. If an emergency shutdown is required, the decision made by the plant manager will be passed to the chief building engineer who, through the technicians listed above, will effect electrical and other shutdowns.

2. The chief building engineer will direct the efforts of other available technicians in the building in handling specific emergencies such as fighting a fire, rescuing a co-worker, tripping off electrical panels, or some other emergency. He will advise technicians of the various departments where they are to trip off or shut off electricity or other services within their particular area of responsibility. He will also advise the plant manager of the progress and the problems he is encountering in effecting shutdown as ordered.

3. All technicians assigned above will communicate with the chief building engineer prior to actually shutting down any major electrical currents, water in the automatic sprinkler risers, or other processes or utilities within their area of responsibility.

C. Responsibilities. Basically, all technicians assigned to the different areas will be responsible for controlling or shutting down process equipment, etc., in their areas during an emergency and for handling the emergency equipment itself. Specific and primary responsibilities for the emergency team are as follows:

1. Chief Building Engineer

    a. Disseminates the plant manager's decision on whether the plant must shut down and the extent of the shutdown
    b. Coordinates the activities of all technicians

c. Performs the shutdown of his area if required

d. Notifies the chief of security of the nature of the emergency and actions being taken

e. Authorizes outside assistance which may be called in by the chief of security.

2. Chief of the Security Force

a. Makes telephone calls to request outside assistance

b. Communicates the nature of the emergency to local management personnel

c. Administers first aid as required

d. Monitors the main gate and office entrance to prevent the unauthorized entry of personnel

e. Keeps records of people entering and leaving the plant

f. Secures additional security personnel as required

g. Insures that all emergency routes to the plant remain open

h. Insures that employees who have been evacuated from the plant do not impede emergency operations

i. Assigns one security guard, if available, as runner to the plant manager's location

j. Insures that the radio communication network of the security force is being fully utilized

k. Supervises to insure the security force is maintaining a log of the events.

3. Technician—Maintenance Department

a. Performs a shutdown of his area if required

b. Assists in handling the emergency.

The list of specific duties continues in this manner, and the responsibilities of each individual listed in Section III A, "Emergency Team," are outlined the same way under the specific team member headings.

IV. Reporting and Taking Action

A. Reporting. All company personnel are responsible for reporting emergencies to their department head or the telephone operator, if an emergency occurs during their shift. The reporting procedure should be brief, as follows:

1. Use interplant phone and call switchboard operator or the department head.
2. Identify yourself.
3. State the nature and the location of the emergency.

The telephone operator upon receipt of an emergency call will immediately telephone the plant manager or the production manager and notify him of the nature and location of the emergency.

If the emergency should occur during second shift operations, the person discovering the emergency will notify the individual in charge of the entire second shift operation, who will then assume the responsibilities as outlined for the plant manager in previous sections.

B. Action. The plant manager, upon receiving word of the emergency, will take the following action:

1. Cause the emergency to be announced over the public address system
2. Inform the chief of the security force of the emergency and direct him to contact whatever outside assistance is needed.
3. Make a decision
   a. Whether or not the emergency can be handled without a shutdown of the plant
   b. Whether or not a shutdown of electric power or fuel is required
   c. Whether or not a shutdown of the entire plant power should be made without evacuation
   d. Whether or not the entire plant should be evacuated
   e. Whether or not certain departments in the plant should be evacuated.
4. Proceed according to the plan required by the decision made and sound the general alarm so each member of the emergency organization and the emergency team will begin to accomplish their assigned emergency responsibilities.

C. Emergency alarm. At the present time, the only emergency alarm system available is the public address system, which, if operated properly, can be heard throughout the entire plant. It may be activated by using one of the microphones located at any one of given positions:

1. The switchboard operator's position
2. The production office
3. The warehouse office.

It may be necessary to call one of these three locations on the interplant phone and have the emergency announced by another individual. In this event, caution must be exercised to insure that the individual who will make the announcement has the message correct. It may be desirable to instruct the individual immediately available to move to the nearest microphone and announce the emergency from a written message prepared for him.

A telephone report must be made to the chief of the security force, whose responsibilities entail securing outside assistance and insuring that emergency vehicles have immediate access to the stricken area when they arrive.

V. Communications

A. Outside assistance. In the event of an emergency, when the plant manager is present, he will call the chief of the security force to inform him of the nature of the emergency and to authorize any outside assistance which may be required. The chief of the security force will be responsible for making whatever outside calls are necessary to obtain this assistance. A list of the organizations which provide fire fighting help, ambulance service, traffic control, and medical assistance will be maintained in the chief of the security force's office and will be kept up to date by the chief. A copy of this list will also be maintained in the plant manager's office and in the chief building engineer's office.

B. Department managers. In the event the emergency occurs in the plant after normal working hours, or after the second shift has ceased operations, the chief of security force will be responsible for calling three management personnel: the production manager on call, the chief building engineer, and the assistant personnel manager.

Because any one of the above may be out of town, his calling attempts will be made in the following order: (List here, in order, the individuals who should be called in sequence until one individual is located at his home.)

The chief of the security force, having made contact with these three management personnel, should keep his outside lines open for calls to request any additional assistance and for important incoming calls.

C. Plant personnel. The plant manager or department managers who have been contacted by the chief of security will estimate the number of men needed to handle the emergency and will have them called immediately. Any special instructions necessitated by the emergency will be given during the call.

The department manager or other individual who is contacted by the chief of security will be responsible for calling the plant manager and the remaining department heads (or their alternates) in reverse order of the above list. Each department manager will be responsible for calling personnel in his department who will be needed to take action during and after the emergency. Telephone lists of all plant personnel will be kept up to date quarterly and distributed by the personnel department.

Because all outside telephone lines will be required for emergency communications, plant personnel are urged not to call the plant unless their call is absolutely necessary.

D. Corporate management. In order to inform corporate headquarters of an emergency, it will be the responsibility of the plant manager (or his alternate) to call a designated official at the corporate level. If the designated individual official cannot be reached, the official at the next corporate level will be called. This procedure will be used until a corporate official has been contacted. During the initial stages of the emergency, the corporate official contacted will be relied upon to communicate information concerning the emergency to other appropriate corporate officials.

(11)

VI. Shutdown Procedures

(These are examples only—develop shutdowns to suit the local situation.)

Two types of shutdown may be needed in an emergency. First, a partial shutdown of the electrical circuits servicing some departments may be required, leaving the remaining departments in operation and, second, a total plant shutdown may be required. In most cases, the first type of shutdown would be put into effect, because it would leave the plant utilities of light, water, air, and steam in operation. With these utilities in operation, plant personnel can handle an emergency more effectively. In either case if procedures in this manual are followed and a training program completed, all areas of the plant should be shut down in less than five minutes.

Gas Release Shutdown. In the event a gas has accidentally escaped from a neighboring chemical plant, shutdown procedures will involve only the shutting down of the plant ventilating system. All exhaust fans in the ventilating system will remain on. All doors which are open for ventilation purposes or for operational reasons will be closed immediately.

Employees should not evacuate the plant when the gas escape alert is given, unless the gas has penetrated inside the plant.

Outline specific instructions pertaining to similar emergencies which may occur locally.

VII. Evacuation

Depending upon the nature of an emergency, personnel may be required to withdraw from certain areas of the plant or from the plant entirely.

A. Procedure. During normal working hours, each department head (or his alternate) is responsible for evaluating conditions at the onset of and during an emergency. If the emergency creates an immediate or potential danger to employees, those who are not required to control the emergency will be directed to leave the plant area and walk to a given assembly area as quickly as possible.

The overall decision of whether or not employees in the assembly areas may leave and depart from the facility with their automobiles will be made by the plant manager (or his alternate) and announced by the chief of the security force.

All employees who leave the facility must check out through the security control at the gate. The security guard will record all departures from the plant during an emergency. When an escape of gas is the reason for withdrawal, employees who leave the plant proper through the main gate should not be required to leave their names with the security guard. However, whenever possible and if it is safe to do so, all employees should be required to pick up their time cards as they depart the plant and leave the time card with the security guard.

If employees are properly oriented and informed, the evacuation caused by escaping gas should be orderly enough to insure that all plant personnel can be accounted for in this manner.

B. Responsibility in the case of gas escape. If the plant must be evacuated because of the presence of gas, it is the responsibility of the plant manager to perform the following:

1. Notify all department heads, by the use of the public address system or by runners, that evacuation must be effected.
2. Double check all phases of the total plant shutdown.
3. Proceed to the main gate and make certain with the chief of security that all employees have left the plant.
4. Lock the plant and leave the chief of security and the security force in charge, since they will be the only individuals in possession of respirators or masks.

C. Plant exits. The following exits may be used to leave the plant building. These exits are not the established emergency exits but can be used during emergency evacuations.

1. The exit from the vault room near the southwest corner of the office building
2. The exit near the southwest corner of the plant between Section 2 and Section 3.

(Continue listing all available exits which DO NOT lead into outside CONFINED areas. The two given are typical examples only.)

Normally, personnel will be expected to leave the plant proper through the main gate where they will be checked out by security force personnel. In certain situations, it may be necessary that employees evacuate the plant property through railroad gates. If the emergency is such that this action is required, the plant manager will be responsible for notifying the chief of the security force, who will have these gates unlocked and opened immediately. These gates should be secured by the security force after evacuation has been effected.

Any doors which are opened or unlocked and are normally used for operational purposes should be used during an emergency whenever necessary. Regardless of what exits were used to evacuate the plant building, employees must get their names and identity numbers to the security guard on duty at the main gate or call the chief of security as soon as possible so all departures can be recorded.

D. Community relations. If it appears an emergency will endanger industrial or nonindustrial neighbors, the chief of security will be responsible for making whatever calls are necessary to alert these neighbors, once the decision to sound the alert is made by the plant manager. A telephone list of these neighbors located immediately adjacent to the plant will be maintained in the security chief's office and will be kept up to date by the security chief.

VIII. Public Information

The essential facts concerning an emergency sustained by the plant will be communicated to the community as rapidly as possible. Families of employees seriously injured or dead will be notified promptly and in person by management personnel before such information is disseminated to the news media.

It will be the responsibility of the plant manager or his alternate, and the industrial relations manager, to act as company spokesmen, and these individuals are the only ones authorized to make any statement to the press. The spokesman will call radio and television stations and the press. Initial information to news media should be patterned as follows:

There has been (nature of emergency; i.e., explosion, fire, release of gas from neighboring plant, etc.) at XYZ Manufacturing Company (report the absence of injuries or number and extent of seriousness of injuries, if this is known.) The extent of damage has not yet been determined. As soon as additional information is available, it will be passed on to you.

If reporters and photographers arrive at the plant, they will be escorted to an appropriate "pressroom," set up near the main office entrance and will remain there until information is released or authorization to take pictures is given. Typewriters and telephones will be made available to the reporters and photographers, if this is appropriate. Photographers and reporters will be treated courteously. However, they should be firmly requested to keep away from dangerous areas and to refrain from interviewing or photographing injured employees. Only verified information will be released.

If the situation dictates, the plant manager and/or the industrial relations manager should prepare written information regarding the actual facts as soon as possible and make it available for immediate release.

IX. Outside Assistance

When an emergency is of such a nature that outside assistance is required, the following agencies, with whom coordination has already been effected, will render service as indicated:

A. Traffic control. Highway: The city police department and/or county sheriff's department will be contacted (*phone number*) _____where traffic control is required, giving the reason for the request.

B. Fire fighting. The city fire department is available on 24-hour call to assist in fire fighting, and the local firehouse telephone number is _____ .

C. Hospital and ambulance service.
1. _____ Hospital will be alerted in the event emergency room service is required. The telephone operator at

_____ Hospital will also assist in procuring ambulance service.

2. _____ Ambulance Service will be called and notified of the emergency and the number of vehicles required will be requested.

3. During first shift operations, the nurse on duty in the plant will be responsible for notifying individual doctors of the emergency and also requesting their presence. During second shift operations, if necessary, this responsibility remains with the security force members.

D. Railroad. The _____ Railroad has a switch engine available in the vicinity. By calling the yard office telephone number, _____ , an engine will be made available immediately. The yard office will also clear railroad personnel from the area of emergency.

E. General. In the event of gas leaks, information regarding wind direction and velocity is available through the municipal airport weather station. The telephone number is _____ .

X. Medical Service

A. First aid. A first-aid room is available in the plant for any personnel who may require first aid during an emergency. During the second shift operations or after normal operations, personnel requiring first aid will report to the security guard for transportation to a medical facility, if the injury is such it cannot be treated in the first-aid room in the plant.

The first-aid room is equipped with the following items which may be required during an emergency:

List only those major items which may be used or required if an emergency should occur.

B. Ambulance. If an ambulance or ambulances are needed to evacuate injured personnel from the premises, the following ambulance and/or medical facilities will be called by a member of the security force during second shift operations or periods when only a few employees are in the plant, or by the nurse, in the event the service is required during the first shift operations when the nurse is on duty.

List by name, address, and telephone number the ambulance services available in the vicinity of the plant.

XI. Equipment and Facilities

List all emergency equipment available for use during an emergency: hoses, shovels, each fire extinguisher, etc., BY EXACT LOCATION.

Facilities available would include buildings used for temporary storage, truck companies to be called to remove salvaged material, etc.

XII. Training

All personnel assigned emergency plan duties will be prepared for emergencies through the following orientation and training activities:

A. Reading. Copies of the company emergency plan are distributed to all personnel who have been assigned emergency responsibilities and duties, and must be read by all.

B. Group instruction. Personnel will be instructed and oriented periodically on the emergency plan during regularly scheduled meetings. This group instruction will be conducted by designated individuals, who will be responsible to the plant manager to insure that all personnel instructed and oriented understand their responsibilities and duties.

C. Special training. Selected personnel assigned to the emergency teams will receive special training from time to time on rescue techniques, fire fighting, and first aid. In addition, these personnel will be given orientation tours of the entire plant to insure that all personnel are familiar with the electrical panels, riser location, emergency fuel shut-off valves, and other emergency equipment which they may be required to operate in the event an emergency occurs.

D. Practice drills. Practice evacuation drills will be scheduled periodically to acquaint employees with their responsibilities under this plan and to develop and maintain the proficiencies of speedy evacuation of the entire building. Practice drills will be held by departments at least semiannually and initially will be gone through step-by-step to insure that all employees assigned throughout the entire building are aware of and know the locations of the established emergency exits they are to use once the evacuation order has been given.

E. Evacuation routes and exits. All management and supervisory personnel are responsible to observe on a daily basis the emergency evacuation routes and emergency exits to insure that these routes and exits do not become blocked through the operational processes.

Blockage of emergency evacuation routes and exits, even for short periods of time, could be disastrous in effecting evacuation. The security force, during their tours, is responsible for reporting all blocked exits and evacuation aisles.

F. Security forces. A copy of this emergency manual will be made available to the chief of the security force, who is responsible to insure that each member of the security force is thoroughly familiar with his assigned duties during emergencies. In addition to this manual, special instructions in written form may be issued to the security force which will be adhered to in implementing emergency action which may be required when the plant is shut down under an emergency. Any changes required in these instructions will be received from the personnel manager, who is coordinator between plant management and the security force. The special instructions issued to the security force will be reviewed at least semiannually and updated where required.

XII. Miscellaneous Emergency Procedures

In the event of snow or inclement weather which may produce hazardous driving conditions, the decision of whether or not the shift on duty will be dismissed early or whether or not the first shift will be required to work the following day will be the responsibility of the plant manager. If conditions are such that the plant manager makes a decision not to have the first shift report for work, plant employees will be notified by an announcement over radio station_____in sufficient time to insure none have yet left their homes to report for work. The responsibility of calling the radio station rests with the plant manager and/or his alternate.

The security force is not authorized to call the radio station to request that a "no work" announcement be made. However, in the event serious or potentially serious inclement weather occurs after normal operations have ceased at the plant, the security force is responsible for advising the plant manager of the conditions in sufficient time to enable him to formulate his decision and effect any necessary announcement.

CHAPTER **17**

Planning for the Bomb Threat

The industrial security program can hardly be judged complete unless plans include the handling of a bomb threat should it occur. The probability of such a threat occurring is dependent upon many factors. The locality of the facility and the type of product being manufactured are probably the two principal elements which should be considered in establishing the potential for such a threat.

The following is a quote from an anarchist, which appeared in several publications in 1972:

> Explosives are one of the greatest tools any liberation movement can have. Ninety percent of all sabotage is based on some type of demolitions or booby traps. The actual application of explosives can be a really thrilling and satisfying experience. Coupled with the destruction of an object of popular hatred it can become more than just a chemical reaction. It can take the shape of hope for an oppressed people.

A publication by William Powell, *The Anarchist Cookbook,* brought out by Lyle Stuart, Inc., 1971, contains this paragraph in its introduction:

> If I could come out in this book and advocate complete revolution and the violent overthrow of the United States of America without being thrown in jail, I would not have written *The Anarchist Cookbook* and there would be no need for it.

The book contains approximately 150 pages with numerous illustrations. It deals with subjects ranging from how to grow marijuana and prepare it for human consumption to the construction of booby traps and bombs. The first chapter contains recipes. One of them, called "Pot Soup," includes in its ingredients "3 tablespoons of grass." Chilly Bean Pot requires "1 cup chopped grass." Other chapters deal with the use of natural, nonlethal and lethal weapons, while others contain formulas for manufacturing explosives out of commonly used household detergents.

Although this publication may not be readily available at every book shop or newsstand it indicates the kind of anarchistic material that can be obtained by anyone wanting it. With such writings available, which may be and often are translated into action, every industrial facility, regardless of what it manufactures or what the process is, must prepare a plan to defend itself against the anarchist, or, if you will, irresponsible dissident.

If an effective physical security program already exists and positive control of personnel and vehicles has been established, a great deal has been accomplished towards the preventive and protective response to the bomb threat. There are three terms commonly used when discussing security against this type of threat.

Bomb threat. A bomb threat consists merely of a telephone call indicating a bomb has been planted in a particular building or on the premises. Usually one may assume that if a person makes a bomb threat by telephone, the caller has some definite knowledge or some definite reason to believe that some type of explosive or incendiary device has been placed on the premises which can cause injury or death to personnel or damage to property. One may also assume that the caller desires to create an atmosphere of anxiety and panic which may result in the disruption of normal activities and routines. Obviously, this would terminate in the loss of production, which very well could be a costly exercise in futility.

Bomb warning. When this phrase is used, it means that a suspect item has been located and there is a reason to believe that it is an explosive or incendiary device, since it closely fits in with the bomb threat which had previously been received.

Bomb incident. The term "bomb incident" means that an actual suspect item has been determined to be an explosive or incendiary device by a bomb technician. The phrase does not mean that it has already gone off, or that it will be detonated.   It does mean that it has been located and has been determined to be a device which could create a major fire or a damaging blast.

Data relative to this new threat are gathered by and are available from the National Bomb Data Center, which is operated by the International Association of Chiefs of Police.[1] The National Bomb Data Center reports statistics on an annual basis from July through June of the following year.

The discussion in this chapter will be limited to bomb threat and bomb incident planning only, because all of the security measures already discussed in the book need no further planning to have their effect in helping to protect the facility against this specific threat.

Bomb Threats

Most bomb threat calls are very brief. The caller will normally state his or her message in a few words and hang up immediately. However, every effort must be made to obtain some detailed information from the caller.

It is of utmost importance that telephone or telephone/switchboard operators are trained to cope with the bomb threat call. Without training, the telephone operator is more likely than not to become overly excited. Training in this subject has the same effect as training in any other type of subject; all training instills confidence through the fact that the individual possesses the knowledge of exactly the actions which must be taken when certain incidents occur.

The telephone operator must not only be trained in the actions which must be initiated when a call arrives, but must be furnished with an easy-to-follow form with a logical sequence to handle and record information. Familiarity with any form on any subject is attained through repetitious completion of the form involved. The checklist form (Fig. 17-1) is very suitable for training the telephone

---

[1] International Association of Chiefs of Police, The National Bomb Data Center, 11 First-field Road, Gaithersburg, Md. 20760.

TELEPHONE PROCEDURES
BOMB THREAT CHECKLIST

INSTRUCTIONS: BE CALM. BE COURTEOUS. LISTEN, DO NOT INTERRUPT THE CALLER. NOTIFY SUPERVISOR/SECURITY OFFICER BY PRE-ARRANGED SIGNAL WHILE CALLER IS ON LINE.

NAME OF OPERATOR_____ TIME_____ DATE_____

CALLER'S IDENTITY
SEX: Male_____ Female_____ Adult_____ Juvenile_____ APPROXIMATE AGE _____Years

ORIGIN OF CALL
Local_____ Long Distance_____ Booth_____ Internal (From within bldg?)_____
                                                        If internal, leave plug in board.

VOICE CHARACTERISTICS		SPEECH		LANGUAGE	
___Loud	___Soft	___Fast	___Slow	___Excellent	___Good
___High Pitch	___Deep	___Distinct	___Distorted	___Fair	___Poor
___Raspy	___Pleasant	___Stutter	___Nasal	___Foul	___ ___
___Intoxicated	___ ___ Other	___Slurred	___Lisp		Other
			Other		

ACCENT		MANNER		BACKGROUND NOISES	
___Local	___Not Local	___Calm	___Angry	___Factory Machines	___Trains
___Foreign	Region	___Rational	___Irrational		___Animals
___Race		___Coherent	___Incoherent	___Bedlam	___Quiet
		___Deliberate	___Emotional	___Music	___Voices
		___Righteous	___Laughing	___Office Machines	___Airplanes
				___Mixed	___Party Atmosphere
				___Street Traffic	

BOMB FACTS

PRETEND DIFFICULTY WITH HEARING • KEEP CALLER TALKING • IF CALLER SEEMS AGREEABLE TO FURTHER CONVERSATION, ASK QUESTIONS LIKE:

When will it go off? Certain Hour_____ Time Remaining_____
Where is it located? Building_____ Area_____
What kind of bomb?_____ **Where are you now?**_____
How do you know so much about the bomb?_____
What is your name and address?_____
If building is occupied, inform caller that detonation could cause injury or death.

Did Caller appear familiar with plant or building by his description of the bomb location? Write out the message in its entirety and any other comments on a separate sheet of paper and attach to this checklist.

ACTION TO TAKE IMMEDIATELY AFTER CALL

Notify your supervisor/security officer as instructed. Talk to no one other than instructed by your supervisor/security officer.

Courtesy of Guardsmark, Inc.

Fig. 17-1. The format of this checklist has been found to be effective for use in the event of a telephoned bomb threat.

operator and recording as much data as the operator can obtain when an actual threat is received.[2]

Training the telephone operator may include members of management phoning test threats to the operators from various locations. Caution should be exercised that the operator has knowledge that test threats will be made and each call should start off with an announcement that may be, "This is a test threat call." During the training period, the calls may be frequent, and once operators have been so inculcated with the desired confidence that they are able to adequately record the substance of these calls, the training calls should be made more infrequently. All test calls should terminate with the operator notifying the security director or the member of management who has been designated as the bomb threat coordinator.

Bomb Incident Planning

Bomb incident planning involves at least seven areas, if the plan is expected to be complete.

Control. The first phase of planning to cope with the bomb incident should include detailed plans which cover at least these subjects:

1. Who will be in charge of the incident?
2. Where will the command center be located?
3. How will critical decisions be made?
4. Who will man the command center?
5. What primary and alternate communications systems will be employed during the incident?
6. What reporting system will be in effect to insure that information is promptly transmitted to the command center?
7. Who will effect necessary coordination with other public or private agencies?
8. Who will deal with the news media representatives?

---

[2]The bomb threat telephone checklist, Fig. 17-1, is taken from the pamphlet entitled *Planning for the Bomb Threat* (Memphis, Tennessee: Guardsmark, Inc., 1971).

Initiation. Plans must include the procedures to be followed upon receipt of a bomb threat or a notice that a device has actually been found.

Evacuation. Decisions must be made as to what procedures will be followed if an evacuation is ordered. In planning the evacuation, consideration must be given to the subject of search of the building or premises and whether or not the search will be conducted with or without evacuation.

The planning of the evacuation and search are closely related, since the search would normally start with the evacuation routes to be taken. This procedure is used because frequently the person planting the bomb knows the evacuation routes and, if he or she is intent upon destruction of life as well as property, the device may well have been placed along the evacuation route and timed to detonate a few minutes after the call was made.

Whether or not to evacuate must be a top command decision, since to evacuate may be costly. The evacuation plan used in the bomb threat incident should be the same as the evacuation plan which would be used during any other emergency. Caution should be exercised, however, regarding the announcement for the evacuation. Consideration should be given to the fact that if a bomb threat evacuation is announced, panic may well ensue. Therefore, employees must understand the entire bomb threat plan to insure not only orderly evacuation if it is necessary, but to reduce the possibility of panic.

Search. The search section of the bomb incident plan must include in detail such items as: what will be searched, what search techniques will be employed, and who will be involved in the searches. It is vitally important that this portion of the plan include sufficient training of those individuals who will be charged with the search, to insure that when the search is placed into effect, it can be accomplished readily yet with absolute assurance that the search of any given area or areas is complete.

To assist in making a decision on who will be trained and charged with the responsibility of conducting a bomb search, the following search systems may be considered:

1. *Search by supervisors.* Using supervisors to conduct a COV-
   ERT search is probably the best method. However, the
   thoroughness of searches conducted by supervisors is normally
   poor and, if the search is detected, an adverse effect on
   morale may occur.

   The advantages of conducting the search through the use of
   supervisors is that the search can be made fairly rapidly with
   little loss of production time, since only supervisors are
   involved.

   The disadvantages are that supervisors are unfamiliar with
   many areas and will usually not look into inconvenient places.
   The use of supervisors generally results in search of obvious
   areas but in inadequate search of hard-to-reach areas. The
   question of violation of privacy and the possibility of danger
   to the unevacuated workers must be considered.

   The thoroughness of a search by supervisors may be calculated at between 50 to 65 percent of the areas being searched
   having actually been searched thoroughly.

2. *Search by occupants.* Search by occupants of the actual area
   being subjected to the search will usually result in a rapid
   search being made. The thoroughness with which the search
   is conducted is somewhat better than that conducted by supervisors, and this system of searching is better for morale, provided training has been given beforehand to instill some
   confidence in the employees.

   The advantages of search by occupants is that the search
   can be conducted rapidly. There is no problem with privacy
   violations to others, and the loss of work time is for a shorter
   period than if an evacuation had been effected.

   Another advantage is that the personal concern for the
   employees' own safety will normally result in a good search
   being conducted. This type of search also has the advantage
   that the personnel conducting the search area in all probability
   those who are the most familiar with the area.

   The disadvantages to the occupants' search is that it requires
   training of an entire work force in any given area. It is also

time-consuming, since it requires that several practical training exercises be conducted. When considering search by occupants, the question of danger to unevacuated workers must be resolved.

The thoroughness of a search conducted by occupants of the area is approximately 80 to 90 percent effective.

3. *Search by trained teams.* This is the best system to use insofar as safety, thoroughness, and employee morale are concerned. However, when considering the economics of the search, this system rates high on the lost-work time scale.

The advantage of the trained team search is that the search is thorough, and there is little or no danger to workers who have been evacuated.

Morale of the employee population is also affected affirmatively, since the employees will usually feel that the company cares for their safety.

The disadvantage of this system of search is the excessive loss of production time, and, since only a few individuals make up the teams, the search is very slow. The trained team system requires comprehensive training and many practical exercises.

As previously stated, this type of search is very thorough and can be calculated as between 90 to 100 percent effective.

Damage Control. Plans must include what damage control measures will be taken and who will initiate the damage control measures.

Disposal. Plans must be made to include how the suspected devices will be dealt with; that is, who will actually remove the suspect devices from the premises, how these devices will be transported, and where the devices will be taken for detonation or dismantling.

Detonation. Plans must include procedures which will be followed if an explosive or incendiary device detonates without warning while the search is in progress or disposal is being accomplished.

The plans regarding detonation must include the procedure which will be used to obtain and utilize such services as medical, utility, and other required support services.

Bomb Threat

Security assessment and planning for bomb threats require the collection and evaluation of information about recent attacks or incidents in the locality where the facility is situated. The processing and evaluation of this information results in intelligence which is used in at least three essential functions:

1. It will insure that this portion of the security plan, to prevent bomb attacks and actual incidents from occurring, is appropriate and practical for the facility.
2. It will serve as a basis upon which to develop a plan which will be in proportion to the existing or potential threat.
3. It will provide for continuous updating and revision of plans as the atmosphere of the locality or local situations change. Updating and revising plans should include adequate countermeasures against new techniques which have been used in actual bombing activities.

Bomb threat intelligence obviously must be obtained to prepare an adequate defense to meet an attack threat at any given time. This intelligence or background information on bombing trends can be obtained in the local area from law enforcement intelligence units, the news media, neighboring communities, and allied agencies or from other established sources of information. As previously stated, national trends can be obtained from the National Bomb Data Center. Monthly NBDC summary reports are published and provide public safety officials with a chronological description and statistical analysis of bombing incidents. Undoubtedly, the local public safety officials have these reports available.

CHAPTER 18

Preparing for Civil Disorders

Management of every industrial facility must formulate plans to assure the safety of their employees and visitors who may be in their facility when a civil disorder occurs. Advance planning for such an eventuality is essential if the overall security plan is to be considered complete.

The geographical location of the facility and the sociological outlook of the community in the immediate vicinity will determine whether or not such advance planning is even necessary. Perhaps all that is necessary is a plan to activate an emergency gate which would be used as the primary point of ingress and egress during periods of civil strife.

Purpose of the Plan

The purpose of formulating a civil disorder plan is to insure that management can continue to discharge its responsibilities in these four areas:

1. Assure the safety and well-being of all personnel within their facility when a disorder occurs
2. Establish full protection of company property from damage and destruction
3. Insure the continued successful operation of the facility with maximum effectiveness
4. Continue their role in community affairs and help to bring about a peaceful and equitable solution of the problem as soon as possible.

Security of the Plan

The knowledge of the contents of any such plan must be closely guarded. The plan should be restricted only to those individuals or groups who are responsible for formulating policy in connection with the plan and implementing it. Knowledge by unauthorized individuals of the contents of the plan would undoubtedly defeat its effectiveness and could well be the chief cause of precipitating such a disturbance.

The plan itself must be handled as company proprietary information and properly safeguarded. It should not be included as any part of an existing emergency plan, although it may contain similar organization and the action it directs may follow, generally, the same pattern of events.

Planning Committee

Once a decision is reached that a civil disorder plan is necessary, a planning committee should be appointed to establish the basic plan and formulate details. This committee should be given the overall responsibility to implement any advance action that is necessary. It should be given the authority of changing or supplementing the facility's physical security program and providing the required administrative functions necessary to implement the plan.

Ideally, the committee should be staffed by the facility's director of security, facility manager, plant fire chief, personnel manager, industrial relations manager, and public relations manager. If the plant organization has a transportation manager, he must be included. Obviously, the makeup of the committee will vary at different facilities, but it should include a wide cross-section of management personnel in those areas that will be directly affected. Lower echelon management may be included, but it is best merely to notify these personnel of the final decisions and make known to them the portion of the plan in which they will be directly involved.

Preplan Intelligence

Advance knowledge of the scope, time, and possible duration of this type of disturbance is vital for management to cope intelligently with the problem.

Information may be gathered from supervisors and other members of the work force. Every level of management must remain alert during times of stress to gather any indications of planned or incipient disturbance.

It may be possible that the security director can acquire intelligence information through his security force. They should have already been trained to report incidents and activities which may indicate that disorders are developing. The security director may develop contact with individuals in the immediate community who are in the position to provide first-hand information of planned or spontaneous activities which would affect the facility. Caution should be exercised in this method of gathering information because, if the source should be disclosed or uncovered, serious repercussions could occur. The safety of the informant must always be considered.

Any information received from the sources discussed above must be carefully assessed as to its validity. Information of this type should be confirmed by another source, whenever possible, before it is acted upon.

If valid information is developed, it should be passed on to the local law enforcement agencies.

Liaison with Local Officials

Responsibility for maintaining law and order, obviously, rests with the local governmental officials. They are responsible not only for the protection of the citizenry and public property but for private property as well.

The security director, as well as every member of the civil disorder committee, should establish and maintain close liaison with the heads of police, fire, and emergency preparedness organizations. A rapport should be established to insure an efficient, cooperative flow of information between the facility's management and these officials.

Local government officials should be apprised of the completed plan and should be notified immediately whenever the plan is to be implemented.

In some instances, it may be possible and desirable to offer the plant facilities to the police and fire departments for use as their command posts. Any side effect of such occupancy must be carefully analyzed, as the establishment of a command post on the premises could result in an adverse effect upon the safety of personnel and protection of property. On the other hand, this arrangement could very well also be quite beneficial. Often it may mean that securing additional manpower for the protection of the facility will not be necessary. When the duration of disorders involves extended periods of time, the economical benefit may well outweigh the possible disadvantages involved.

Civil disobedience will usually be an extension of some type of demonstration. The disobedience is generally intended to provide the demonstrators with more publicity. It may take the form of sit-ins, passive resistance to arrest, overt violence, or many other forms. Whatever the means used, they will result in violations of the law, and therefore local police authorities become directly involved.

If police organizations and members of the security force become involved physically on or near the company's property, photographic coverage is a must. Company personnel should be assigned to photograph these activities with both still and motion cameras. If any subsequent charges of brutality are made against the police and company security guards these photographic records will help prove the truth of the case.

Physical Security

The portion of the plan which outlines the physical security measures to be implemented should include only those changes to the overall security plan already in effect.

The physical security changes for the civil disorder plan will probably only require that certain decisions already made regarding emergency evacuation plans, disaster plans, and plans to cope with labor disturbances are extracted and tailored to accommodate this plan.

CHAPTER 19

Increasing Protection During Labor Disputes

The necessity may arise for the protection plan to be augmented by additional security measures if management-labor negotiations deteriorate and it appears that a strike by the employees is inevitable.

The security director should immediately review the present protection plan to insure that all phases of the security program are being implemented and executed efficiently and effectively.

He should also review appropirate sections of the emergency and disaster plan and civil disturbance preparedness plan, if such a plan has been prepared. The fire protection and prevention plan should also be reviewed. Emphasis should be on ascertaining whether the portions of all these plans in which the security force has been assigned certain responsibilities have been updated and still apply.

The plant security director should then personally review with the entire security force the responsibilities outlined in these plans and conduct such necessary training or retraining as may be required to assure himself that every member of the security force can effectively discharge the responsibilities should the occasion arise.

Experience has indicated that in addition to accomplishing those tasks discussed above, additional measures for the protection of the property may be considered. A checklist follows which will be helpful to the security director in accomplishing his mission. It may also prove quite useful in compiling some recommendations to submit to management.

It is important that if any of the measures listed below are put into effect the security director should receive the authority and support of plant management to assist implementation. Consideration should be given to a study of this material when management-labor negotiations are pending, because some of the items may require considerable time to accomplish. Many of the measures can help reduce the likelihood of unfavorable incidents occurring.

1. Change all perimeter gate padlocks and locks on the peripheral doors of all building to which keys have been issued to employees who may join the strike.
2. Recover keys issued to employees who, in all probability, will go on strike.
3. Consider recovery of company identification cards when the employees leave the plant for the last time before the strike, or collect the identification cards when the last check has been issued.
4. Notify the security force of all outstanding identification cards held by striking employees by the bearer's name and card number.
5. Issue special passes or special identification cards to employees who will not go on strike to insure their proper admittance when reporting to work.
6. Notify employees who will continue to work, to keep windows of their automobiles rolled up and doors of cars locked when driving in and out of property that is being picketed.
7. Advise employees not to stop whether on foot or in automobiles, if possible, during ingress and egress of the facility.
8. Caution employees not to be encouraged into rolling their automobile windows down or getting out of their cars, by foul or abusive language directed at them by the strikers.
9. Secure all peripheral doors and gates not being used during the strike, and keep them secured.
10. Have all combustible trash removed from the property inside and outside the facility before the strike takes effect.
11. Cut and remove all underbrush, weeds, etc., along the perimeter barrier (inside and outside) and inside the protected area.

12. Move combustible trash receptacles in outside areas at least 50 feet from any perimeter fence and insure that they have been emptied.
13. Attach outside hoses to yard hydrants for immediate use. Remove all wrenches from post indicator valves outside the protected area.
14. Check all standpipe hoses, fire extinguishers, and other fire-fighting equipment *AFTER STRIKING WORKERS HAVE LEFT THE PLANT.*
15. Test automatic sprinkler and other automatic fire protection systems *AFTER STRIKING WORKERS HAVE LEFT THE PLANT.*
16. If any building roof creates a possible fire hazard from hand-propelled fire bombs, have ladders, hoses, and extinguishers available. It may be desirable to have extinguishers available on the roof.
17. If construction of the roof permits, consider flooding the entire roof with 1½ to 2 inches of water at critical periods of the labor dispute.
18. Cover windows facing the perimeter or be prepared to cover them by using plywood or other suitable material. See Fig. 19-1.
19. Protect company vehicles by parking them well inside the protected property. If no perimeter barrier is erected, consider dispersing them elsewhere.
20. Park vehicles of employees, who are still working, well inside the protected area.
21. Station uniformed security guards on company property, *not* on public property. Define and mark the perimeter of the are to be honored by any pickets.
22. DO NOT ARM security guards and DO NOT ASK security guards to take pictures or tape conversations. These actions will usually only antagonize strikers and turn them against a security force which otherwise would enjoy some semblance of neutrality and do a better protection job.
23. Station security guards or establish additional patrols to insure that all outside areas can be observed effectively.

Fig. 19-1. Several manufacturers of impact-resistant glass offer greater protection than the universal "safety glass" previously relied upon. The same impact on each pane of glass easily defines the tensile strength of the upper (safety) glass and the lower (plastic) which is mar-proof.

24. Replace all lights previously burned out and not replaced and consider installation of additional illumination to eliminate heavily shadowed areas near the perimeter barrier.

25. Establish as few points of ingress and egress to the plant as is consistent with operational efficiency. Secure these points with uniformed security men.

26. If considering employing people to replace striking workers, establish an employment office temporarily elsewhere, such as a motel or a vacant store, and advertise the location in the employment ad. Refrain from expecting, initially at least, employment applicants to negotiate picket lines.

27. Arrange for vehicular shuttle to out-of-the-way bus stops for those employees who will continue to work and use public transportation. Advise working employees not to use bus stops which are relatively close to lines or groups of pickets.
28. Consider a temporary shuttle service to and from nonstriking employees' homes or areas, if necessary.
29. Compile a list of names and telephone numbers of all law enforcement officials, the fire department, and other key personnel.
30. Establish and maintain close liaison with local police authorities and leadership of the servicing fire department.
31. Keep cafeteria or vending type lunch machines operating for the convenience of working employees. Make this type of arrangement ahead of time. It may be necessary to provide additional stocks for vending machines, if the vending machine servicemen will not pass the picket lines of striking workers.
32. *Call* in advance all vendors normally servicing the facility to determine whether or not truck drivers of their companies will continue to service the plant during the strike. Provide for diversion of incoming shipments, if necessary, to locations outside the disturbed area.
33. Review procedures pertaining to annoyance and threatening calls with telephone operator.

CHAPTER 20

The Employee Security Education Program

Any security protection plan designed to eliminate security hazards will prove difficult to administer unless it is supported by an effective employee security education program. It is most difficult for security personnel to effectively accomplish their mission unless they have the active interest and support of the majority of the employees in the facility. This support can most easily be obtained through an effective security education program.

The goal of such a program is to acquaint all employees with the reason for the security measures being planned and why they will be implemented. The plant security director, with the support of top plant management, will be responsible for implementing the education program. The objective of the program is to inculcate all employees with the need for the program and instill in them a willingness to assist in and comply with the plan.

Security consciousness is not considered an inherent state of mind. This attitude must be acquired. Many people, not excluding top management, are naive and trusting and are inclined to accept situations at their face value. Numerous instances can be recalled; at the onset of an investigation at an industrial facility, top management least suspected "good old John, he's been with the company for eighteen years." The investigation later revealed that "old John" was, indeed, the brains directing the thievery, simply because, being trusted, he had latitude to operate and also possessed intimate knowledge of the facility.

If the industrial facility produces and distributes an in-house news letter, this media is an effective means of reaching all employees. The planned security program can be presented in its entirety in one such issue or can be presented in several parts. One vehicle which has been used with considerable success in promoting the educational program is a form in the in-house organ soliciting suggestions from employees on how the security in their particular area may be effectively increased. The results are often enlightening and rewarding.

Another method that may be used is placing announcements pertaining to the program in the paycheck envelopes or disseminating this information through supervisors, safety committee members, or local union leadership. The method employed is only limited by the ingenuity and resourcefulness of the security director.

Past experience has proved that some employees resist firm establishment of a security program for no apparent reason other than the assumption that they are no longer to be trusted. This, obviously, is not the reason for establishing the program, but rather the opposite. The establishment of physical security assures the employee that no one unauthorized is allowed to enter the facility and possibly commit acts of theft, sabotage, or vandalism. If significant acts of sabotage or vandalism occur they may have an influence on the employee's expected work performance or could, for example, result in the employee being laid off temporarily while repairs are in progress. Losses through theft obviously have an effect on the company's ability to make a profit, which, in the long run, influences the ability to increase wages and salaries and provide various employee benefits. This fact should also encourage employees to support security measures.

If the proper educational program is launched, the reason for the program being restrictive made clear to the employees, and this information disseminated to all employees, the establishment of the entire plan will be accepted with little or no objection from the employees as a group.

The effective employee security education program must first be outlined in its entirety, and then step-by-step procedures established as to how it can best be implemented; keep in mind, however, that there will always be several items and procedures in the protection plan which are best not revealed to the general employee population.

CHAPTER 21

Organizing the Security Guard Force

Readers will recall that in the opening paragraphs of the book a statement was made that caution should be exercised in attempting to fill gaps in the protection plan by increasing the strength of the security force. This statement was followed by another explaining that in order to determine the strength, organization, and deployment of the security force, a complete physical security survey must first be accomplished.

Only after the study and analysis of existing conditions has been completed and the degree of security required throughout all areas determined can an intelligent decision be made relative to the size, organization, and deployment of the security force which will ultimately execute the protection plan.

This should now be readily apparent, since the security of the entire facility will undoubtedly be upgraded by the installation of various physical aids to security, the formulating of new policies and procedures, and numerous other improvements incorporated into the protection plan. The objective should be "maximum security at minimum cost."

The next decision that must be made is to analyze whether or not to employ in-house guards or have the security program administered through a contract security company. The question is who will furnish the necessary manpower, training, supervision, and expertise to execute the plan?

It also follows that consideration should be given to providing a reserve of trained replacement personnel to cover periods of sickness, vacations, and terminations. Someone must be appointed who will be responsible to continually re-evaluate the protection plan as physical and operational changes occur within the facility.

Contract Guards versus In-House Guards

If a guard force is already deployed at the facility, with personnel on the company payroll, a decision must be made as to the disposition of these men. Are they of age to retire? Can a percentage be retired and others integrated into the work force? Are some of them qualified to be absorbed in the ranks of a contract security company to be employed elsewhere? Or, possibly, should they be just let go? Do the answers to these questions rule out changing to a contract security company?

In order to arrive at an intelligent decision about the disposition of these men, a comparison of the advantages and disadvantages of in-house guards versus contract security guards must be made. In the examination of these two sources of manpower, keeping in mind the economics involved, and, more importantly, the effectiveness of the security force, the following discussion on in-house guards applies whether or not company employees are presently performing the security function.

In-House Guard Force

Economics. The hourly rate of pay of company guards will be directly affected by the wages of a comparable category of employees within the facility. Payroll costs and company fringe benefits must be calculated. If the employees are unionized, what effect will this have on the security force?

Uniforms will most certainly be needed if the security force is to assume any authoritative role in the program. What type of badge will be used? Provisions for inclement weather, clothing, and gear will have to be considered.

Equipment for the force is another expense. Such equipment as explosion-proof flashlights, nightsticks and revolvers, and various report forms such as vehicle logs, gate passes, guard daily reports, and guard log books will have to be purchased and in all probability will require special printing.

Training for the entire force in their primary duties for all posts will have to be accomplished. If weapons are required, ranges and qualified instructors will be needed; first-aid training is a must for every member of the force; government clearances may be required in some instances; guard-force supervisors will require advanced training; and annual retraining programs will have to be initiated.

Company labor disputes or work stoppages may occur. What position will the company guard force assume? How will the security force strength be augmented under these conditions?

Civil disturbances or riots in the immediate vicinity of the facility may well occur. How will the guard force react, particularly if they live in the neighborhood? Again, additional manpower will certainly be required on short notice and probably for only a short period of time. Has the in-house force received any training in crowd control?

Licensing and police commissions are a certainty in the near future. Will the company guards qualify?

Police department and plant security force liaison must be established and maintained. Who will be responsible?

The above are only some of the questions to which answers must be sought. They do represent, however, those areas in which capital expenditure and increased recurring costs are the major considerations.

Contract Security Guard Forces

Consider each item discussed above in the light of some paragraphs from Volume 2 of the Rand Report.

The overall cost of in-house guards could be higher than contract guard fees for a variety of reasons. In-house guard wages are typically higher because they are influenced by in-house, non-security employees whose wages are higher, often because of collective bargaining agreements, and because in-house guards generally have more seniority on the job than contract guards.

Large contract guard firms benefit from economies of scale in hiring, training, insurance and other costs. For temporary or special-event guard service, the costs of procuring and training in-house guards may be prohibitive, even if such temporary in-house personnel can be recruited in sufficient numbers.

Administrative unburdening. Hiring contract guards relieves the client of the need to develop and administer security personnel recruitment, screening and training programs. It also relieves him of the need to provide close supervision and special liability insurance uniforms, and equipment for the individual guards.

Some sources argue that it relieves the client of the need to have a security expert on its management staff. Hiring contract guards solves the administrative problem of scheduling manpower when someone is sick or on vacation, or when additional man-hours of guard services are temporarily required.

Availability of manpower. During periods of illness, unexpected absence, vacations, or peak demands for guard services, it often is necessary to have substitute or supplemental guard employees. In a small in-house guard force, extra guards may not be available, resulting in a lapse of security. If extra in-house guards are available, they may be inefficiently used most of the time. Contract guard firms with larger pools of manpower can use their personnel efficiently, while still having adequate substitute or supplemental guards on short notice. The smaller the guard force needed at one location, the greater the significance of this manpower-availability issue.

In a related vein, if it should be desirable to reduce guard manpower levels, or to eliminate a particularly undesirable individual guard, this is easier to accomplish with a contract guard force.

Unions. The three arguments presented by users of guards in favor of non-union guards are (1) they are less apt to strike; (2) they are less apt to support overtly or sympathize with striking unionized non-guard employees; and (3) non-union employees tend to earn less and have fewer fringe benefits. Since reportedly 90 percent of unionized guards are in-house guards, the above arguments favor hiring contract guards. Even if in-house guards are not unionized, they may benefit from gains made by unionized non-guard employees. Only 10 to 25 percent of the guards employed by the three largest contract guard agencies are unionized.

Impartiality. Contract guards may be more consistent and impartial in enforcing regulations than in-house guards. This is said to be possible because contract guards, having a different employer and relatively low seniority, form fewer close associations with non-guard employees of the client. This issue is especially relevant in cases where in-house guards tend to be long-time employees "pensioned off" to a guard post when incapable of adequately performing their former jobs.

Security expertise. When a client hires contract guards, he also hires the contract guard agency management and its security expertise. This assumes that the contract agency being considered can attract or produce better security experts than an in-house force. This is probably true for a firm in need of a relatively small security force. Also, since contract guard management consists of full-time security men who must continually compete, there may be more incentives for them to stay abreast of the state of the art in security than for an in-house security manager.

Training. The management of the large contract guard firms claim they offer better trained guards than most in-house forces currently utilize. The large contract firms say they can afford to develop a good training program, hire good instructors, and efficiently train their employees because of economies of scale not enjoyed by most in-house forces.[1]

Supervision of a contract security guard force is furnished by the contract security company on a 24-hour-per-day, 7-day-per-week basis. Supervisors must submit written reports of their activities to their management. Management on higher levels in the contract company conduct unscheduled, unannounced inspections of operations they service, completing the supervision plan.

Uniforms and equipment are furnished by the contract security company and are included in the basic hourly rate.

Required licensing or commissioning of guards is also accomplished, and any cost is included in the basic rate. Liaison with local police and sheriff's departments has already been established and will always be maintained.

The problem of reassigning or terminating the company employed guard is most difficult. Obviously, each individual must be handled separately. It may be possible to absorb some of the men into the facility work force without a great remaining cost. Some men may qualify and be employed by the contract service (for assignment elsewhere). Others may already have reached retirement age. There is no clear-cut solution to the problem, and, in some instances, individuals may suffer personal loss. But, for company needs, the qualifications and effectiveness of each individual man must be weighed against the need to establish an effective, efficient overall security force.

---

[1] Sorrel Wildhorn and James S. Kakalik, principal investigators. "The Private Police Industry: Its Nature and Extent," *The Rand Reports*, vol. 2. The Department of Justice, December, 1971.

Technical Considerations

There are other major considerations of a more technical nature which must be resolved and/or should be considered in organizing the functions of the security force, be it in-house or contractor furnished.

When constructing or renovating a guardhouse, the following features should be considered:

1. The structure must be of sufficient size to permit the guard or guards to carry out their functions efficiently. Figure 21-1 shows a portable guardhouse which may be used as a temporary shelter for fixed guard points. In many instances, the addition of heating or air conditioning can convert the shelter to a permanent installation.

Courtesy of J. Henges Enterprises, Inc.

Fig. 21-1. Portable guardhouses may be used as temporary shelters for fixed guard posts. In many instances, the addition of heating or air-conditioning units converts these shelters to permanent installations.

2. The guardhouse should have glass installed (possibly, impact-resistant glass) on all four sides so no blind spots obstruct the security guard's observation.
3. Doors should be of the sliding type rather than hinged, particularly when the structure is installed in the center of a roadway. This will assist in more efficiently controlling vehicular and/or pedestrian traffic.
4. Lighting should be arranged so the illumination during the hours of darkness does not create a glare on the glass, thus restricting vision.
5. If the post is manned on a 24-hour basis, and the organization of the security force does not permit any temporary absence from the post, rest-room facilities will be needed.
6. An overhang should be created outside the structure to shelter the guard and personnel being identified or documented during inclement weather.
7. If an overhang is required and personnel are controlled during the hours of darkness, sufficient illumination should be installed in the overhang to permit identification and documentation of personnel.

Security force vehicles which may be employed in the area security patrol plan should be selected in relation to the conditions that exist. If the routes patrolled are hard surfaced and will not deteriorate during inclement weather, the three-wheeled, gasoline-driven vehicle is an excellent conveyance.

If the patrol routes require negotiating unimproved terrain, a four-wheel drive vehicle should be considered.

When vehicles are employed, consideration should be given to equipping them with portable hand extinguishers of suitable type, first-aid kits, and a powerful hand-operated 12-volt searchlight.

Guard supervisory systems must be installed whenever foot or vehicular patrols are included in the security force functions. There are two methods of supervising these patrols:

1. Electronic supervision, consisting of stations installed throughout the protected area. The stations are electrically connected to a monitoring panel and monitored either at an outside central or proprietary station.

If the electronic supervisory system is monitored anywhere other than on the premises being protected, it will normally require that patrols are started at a specific time and completed within a predetermined period. This dictates that a pattern must be followed and detracts immeasurably from the security value of the patrol.

The greatest advantage of this type of outside system is derived when a single security guard is on duty in a facility, because it insures that his activities are monitored by others.

2. Mechanical supervision systems, which consist of watchclock key stations being installed at critical points along the patrol route. The security guard carries a clock, and when each station is visited the "key" (a recording device) is inserted into the clock and an impression made on a disc or tape which will indicate the location and the exact time the station was visited.

The watchclock is undoubtedly more suited to supervise both foot and vehicular patrols, because it does not require that an outside company be contracted to move stations. The small key stations can be relocated without difficulty by the plant maintenance department. This system is particularly suited to supervise outside area patrols.

The mechanical supervisory system discussed here is far more effective in accomplishing the security mission, because no set pattern as to direction or time need be established. Normal procedure is to permit the security guard to commence his rounds within a 20-minute period which will not restrict him to a specific sequence or specific route.

It is a good practice to consult fire insurance underwriters before designing security guard patrol routes. Quite often it is possible that a reduction in the insurance premium can be accomplished if the tour routes are properly designed and patrolled at a specified frequency.

Security force communications has a direct bearing on the efficiency with which they can operate. Security guards must be able to communicate with each other, not only when an emergency arises but routinely.

Supervisors must have a means of communicating with each fixed post and patrolling or moving guard if they are to be expected to direct the efforts of the force. It becomes increasingly important that the supervisor have constant communication with his security force during periods of emergencies, because the security and safety conditions change rapidly.

Telephone communications must be established between all fixed posts, and telephone communication should be available to the patrolling guards if no other means of communication are available to them. Often when operations are shut down in industrial facilities, only a few telephones remain active. If telephone communication is solely relied upon during inside or outside patrols of the facility, active telephones must be available along the patrol routes and security guards must know the locations of these active telephones.

Ideally, both telephone and radio communications should be furnished the security force. If outside foot or vehicular patrols are conducted, radio communication between the supervisor and the patrols is necessary.

When radio equipment is being selected, consideration should be given to using portable units instead of units installed as base stations or in vehicles because this will increase the mobility of the radio communications network. Portable radios in the vehicle can be placed in battery chargers while the patrolling security guard is in the vehicle. When he dismounts he is able to remain in constant communication by carrying the portable unit with him.

Portable units assigned to fixed posts can remain in chargers and, if necessary, unit rotation arrangements can be set up to insure patrolling guards are provided with fully-charged equipment.

Paging systems may be considered as a secondary means of communication between members of the security force. However, efficiency of operations is reduced because the security guards being signaled must still move to the nearest active telephone.

Audio or public address systems are not suitable as a means of communications between members of the security force, because everyone within hearing distance is aware of the messages being communicated. The use of the public address system will usually pinpoint the guard's location to the employees, when otherwise they may not know his whereabouts. Coded signals are also of little value because the codes, of necessity, must be simple and can soon be broken by the employees.

The Security Force Procedure Manual

A security procedure manual should be written to insure that every member of the security force is familiar with the duties he is expected to perform. The contents of the manual will be dictated by the security program established at each individual facility. To be effective, it must be reviewed frequently and updated. An effective security program requires frequent changes; therefore, it is desirable that the procedure manual should be in the form of a looseleaf notebook so when changes occur, one entire page can be retyped, the obsolete page removed, and the revised instructions inserted in its place.

In general, the format of a security procedure manual may well include some or all of the subjects appearing below.

Purpose and Scope. The purpose of the manual should be clearly outlined. A statement should also appear on how the procedural changes will be incorporated into the manual so it remains an effective day-to-day guide. This section should also contain the facility security force organizational chart.

The facility security policy should be stated and detailed enough so that each security guard understands the consequences should he fail to carry out fully his assigned duties.

Plant Protection. This section will outline all of the specific duties assigned to each fixed and patrol post. The posts should be numbered either 1, 2, and 3; or A, B, and C and so on, and the specific duties assigned each post outlined under these subheadings. Exact hours each post is manned should be indicated. Also included should be a statement of whether or not the particular post is armed. This section may well contain a list of duties that are performed by the guards at each one of the posts; chronologically arranged. These lists are useful as checklists by newly assigned guards and for review by guards with more seniority.

The use of access rosters, information pertaining to facsimile signatures to be used as comparison of signatures on passes, and all those protection duties performed by the security force at the facility should be outlined in this section of the manual. These instructions must be specific and should follow a logical sequence of events from the time the guard reports on shift until he completes his tour of duty.

Plant Protection—Patrols. This section should contain a list of the key stations installed on each patrol route and a description of their specific locations. Instructions should include times that patrols will be conducted and the manner in which the patrol will be implemented. This section may include care of vehicles and/or radio communications if this equipment is being used.

Plant Protection—Miscellaneous. This section will contain such duties as raising and lowering the American flag, escorting groups of visitors through the facility, escorting trash removal trucks to the dumping grounds, inspection of the perimeter barriers, and all those other protection duties assigned to the security force which may be done on a regular or irregular basis.

Emergency Telephone Numbers. All emergency telephone numbers which the security force may be expected to use should appear in the front of the procedure manual to insure ready reference.

Fire. Under this section all instructions pertaining to fire prevention, protection, inspections, action to initiate in the event of fire, automatic sprinkler tests, inspection of fire extinguishers, and any other duties related to the fire protection and prevention plan of the facility should be outlined. To insure that the various subjects can be easily and quickly found, they should be subtitled and alphabetized.

Ambulance Service. Instructions pertaining to the use and operation of the facility's ambulance, if one is available, should appear in this section and if outside ambulance services are relied upon, specific instructions pertaining to the method these services are secured should be outlined.

Hospitalization—First Aid. Instructions relative to the hospitalization of injured employees and the extent of the first aid to be administered to the employees in the facility's first-aid room should be detailed in this section.

Employee Identification System. In this section should appear such items as the procedure to use in ordering badges, action to be taken in the event an employee forgets or loses his identification badge, the type of visitor identification badges in use, construction badges, and any special type of badges. Again, subparagraphing will assist guards in finding specific instructions relating to this subject.

Employee Entrances and Exits. This section should contain specific instructions as they relate to the controls established at each authorized employee entrance and exit and the positions that the security guards are to take to check identification cards, conduct lunchbox or purse inspections, and other duties they will perform at specific points.

Employee Absenteeism. If the security guard force will be responsible for accepting calls from employees who are reporting absent or if they will be responsible for reporting employees who arrive late, these instructions should be included in this section of the manual.

New and Discharged Employees. Instructions relative to the guard force's responsibility in escorting discharged employees from the premises or escorting new employees to their work areas or to their supervisors will be included in this section.

Property Receipts. If sales are made to employees, either of finished products or scrap material, or if the employees will be given any type of property belonging to the company to be carried from the protected area, the exact procedure must be outlined. Copies of property receipts, gate passes, or other such forms used should be included in the security procedure manual as an annex.

Tool Check-Out. Often tools are loaned to employees, and the security force is responsible for establishing a record file to insure that the tools are returned at the expiration of the period loaned. These instructions should show the type of pass required to remove these items and give details of the type of record to be maintained.

Traffic Rule Violations. When the security force is responsible for enforcing traffic regulations and parking regulations, procedures they are to follow must be outlined. Traffic violation tickets, if used, should be included as an annex to the manual.

Inspections–Personnel and Vehicles. If the security force will perform lunchbox, purse, or package inspections, and if they will be required to inspect the cabs of the trucks or any other vehicles including railroad box and tank cars, detailed instructions should cover the exact method they are to use.

Telephone Calls. Often the security force will man the facility's telephone and receive incoming calls after the normal operating day. In these instances, logs should be furnished and instructions on handling the calls put in writing. Copies of the logs that will be used should be included as an annex to the manual.

Vehicles Entering or Departing the Facility. This section should include the procedures to be used in logging vehicles in and out and all other information pertaining to the specific duties involved in controlling specific vehicles or all vehicles in general. Registers or logs involved in these duties should be attached to the manual as annexes.

Visitors. All regulations pertaining to the handling of visitors, including visitor parking stalls and the documentation and handling of the visitors, should be outlined in this section of the manual.

Mail Service. If the security force at any time will be responsible for handling incoming or outgoing mail, specific instructions in the procedures to be used must be included.

Intoxication. Specific instructions must be included relative to the handling of intoxicated employees or employees who appear to be under the influence of drugs.

Law-enforcement Officials. Specific instructions must be issued the guard force concerning the authority of law-enforcement officers entering the property either to arrest or issue subpoenas to individual employees. These instructions should also include the action the guard force should initiate if a finance company employee should arrive to repossess an employee's personal automobile.

Photographic Authorization. Include in this section whether or not cameras are authorized and all restrictions pertaining to photographic activities, which may include activities of the press.

Lost and Found. The security force is normally responsible for the security of personal items found on the property. If they will handle lost and found articles, they should be furnished with a secure storage container and should be given specific instructions pertaining to the recovery of lost items and the actions required of them when lost items have been reported to them.

Key Control. If key control has been assigned to the responsibility of the security force, procedures for issuance, recovery, and maintenance of records must be included in the procedure manual.

Plant Shutdowns. Specific instructions pertaining to the security force's functions during plant shutdowns must be explicit.

Union Activities. The security force must be given specific instructions concerning union representatives calling at a nonunion plant and those applicable instructions or rules they are to enforce regarding union activities by union employees.

Security Force Report Forms. This section should explain the use of each of the forms, when they should be used, how they should be completed, where the supply of forms is secured and the distribution or filing of the completed forms.

All forms listed in this section of the procedure manual should be included as an annex with a sample completed form which can be used as a guide.

Security Force Duties

The duties and functions assigned to each post and/or each member of the security force are many and varied. The exact functions, obviously, will depend upon the program in existence. Any duty related to the security of property and safety of personnel can conceivably be performed by the security force.

At some facilities, the duties may involve some rather technical functions, in which case the security force will require specialized training. Models of industrial facilities may be useful to help simulate security problems, and thus serve as a training aid (see

Fig. 21-2). At other facilities, where no company proprietary or government classified operations are in progress, the functions will not be more involved than those considered routine duties of any security force employed at an industrial facility.

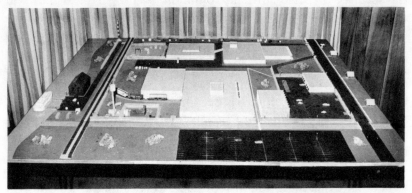

Courtesy of Guardsmark, Inc.

Fig. 21-2. Models of industrial sites are often used by contract security companies to study or simulate security problems which are then analyzed. These models are also used to train the guard force before they arrive at the facility to be protected.

The security force must not be assigned duties of a janitorial nature or any other menial tasks that would detract from the primary mission. Duties should not dimish the authoritative position created for and occupied by this special group of men. Men in the security force should be selected because of their honesty, integrity, stability, and physical attributes. They must also be capable of mature judgment, making timely decisioñs, and be able to discharge their duties firmly, fairly, and with impartiality. They must be capable of performing cooly and precisely during emergencies and be morally beyond reproach.

Bibliography

General Electric Co., Lighting Systems Dept. *Lighting for Safety and Security.* Hendersonville, N.Y.: 1970.

Guardsmark, Inc. *Planning for the Bomb Threat.* Guardsmark, Inc. 22 South Second Street, Memphis, Tenn. 38103: 1971.

Honeywell, Inc. *Planning Guide, Building Security* Form No. 54–0349, and *Planning Guide, Fire Protection* Form No. 54–0362. Honeywell, INc., Inquiry Supervisor, 2701 Fourth Ave. So., Minneapolis, Minn. 55408.

Hunter, George. *How to Defend Yourself, Your Family, and Your Home.* David McKay Co., Inc., 750 Third Ave., New York, N.Y. 10017: 1967.

Illuminating Engineering Society. *American National Standard Practice for Protective Lighting.* RP-10. 345 East 47 Street, New York, N.Y. 10017: June, 1970.

Industrial Fire Brigades. *Training Manual.* NFPA, 470 Atlantic Avenue, Boston, Mass. 02210: 1968.

Industrial Publishing Co. "Guide to Plant Fire Protection." *Occupational Hazards.* Reprints available from *Occupational Hazards,* Cleveland, Ohio: 1964, 1965.

International Association of Chiefs of Police. *Bomb Incident Procedures.* National Bomb Data Center, 11 Firstfield Road, Gaithersburg, Maryland 20760.

National Fire Protection Association. *National Fire Codes,* vols. 1-10. 1972-73. Pocket Editions NFPA Nos. 14, 10A, 80: 1970, and, *Guide to OSHA Fire Protection Regulations,* vols. 1-5. NFPA 470 Atlantic Avenue, Boston, Mass. 02210: August, 1971.

U. S. Dept. of the Army. *Physical Security.* FM 19-30. GPO, Washington, D.C.: November, 1971.

Appendix

This checklist is designed to follow the same sequence as the chapters in this book, and is furnished to assist the security director in conducting the physical survey of his facility. It does not cover Chapters 11 and 12, and 17 through 21, as these chapters do not lend themselves to the checklist procedure.

Administrative Details—Chapters 1 and 2

1. Secure plot plan, aerial photo, etc., of facility.
2. Secure employee rules, handbooks issued, etc.
3. Is there now or will there be new construction, and when will it start?
4. If answer to the above is affirmative, sketch in the area on the plot plan.

A. Employee Strength and Shift Times

1. Total number salaried employees.
2. Total number hourly employees.
3. Normal operational days—S M T W T F S.
4. Increases or decreases in seasonal production and how they will affect security.
5. Times of shifts, breaks, lunches, etc.

   a. 1st shift from ____ to ____ breaks A.M. ____ P.M. ____
      lunch A.M. ____ P.M. ____
   b. 2nd shift from ____ to ____ breaks A.M. ____ P.M. ____
      lunch A.M. ____ P.M. ____

   c. 3rd shift from ____ to ____ breaks A.M . ____ P.M . ____
      lunch A.M . ____ P.M . ____

6. Consider any overlapping in shifts.
7. Are employees authorized to leave the plant during lunch?
8. Are maintenance men regularly in the plant during "down" periods?
9. What is the approximate annual percentage of labor turnover?
10. Percentages of male and female employees.

B. Eating Facilities

1. Location of cafeteria (hot cooked meals).
2. Is it company or concession operated?
3. What are the hours and days of operation?
4. What are the hours of operation before and after the open period?
5. What is the method of receiving supplies and disposing of garbage?
6. How are stored foodstuffs and supplies secured?
7. Are vending-machine type lunchrooms available?
8. Pinpoint each location on the plot plan.
9. Are vending machines located outside lunch areas? Where? Plot location on plans.
10. If changemakers are available, where are they located? Plot location on plans.
11. How are cash funds handled? How much daily?
12. Are employees allowed to eat lunch at their work stations?
13. What control should be exercised over concession employee personnel?

C. Custodial Service

1. Are they contract or company employees?
2. Hour work starts. Finishes.
3. What procedures are used to control custodial employees?
4. Number of men. Women.
5. Are supervisors present?

6. How do employees get into facility?
7. Do guards sign them in and out?
8. How is trash removed from each building or area they work in?
9. Is trash taken outside the building? What type storage container?
10. Do contract personnel bring their own equipment and supplies?
11. If contract, where are their vehicles parked?

D. Company Sales Store

1. Where is physical location within facility? Plot location on plan.
2. What days and hours is the store open?
3. Are sales made to the public?
4. How are sales recorded and paid for?
5. Study in detail how an employee would make a purchase.
6. How many people work in the store?
7. Break down duties of each of the above.
8. Does procedure call for self-service?
9. How is stock obtained from finished goods warehouse?
10. How is stock accepted and by whom?
11. How often and by whom are inventories taken?
12. What type of packaging is used?
13. Can employees take packages to their lockers or work areas?
14. How are cash sales handled and secured?
15. How is trash from the store removed?
16. Who cleans the store?
17. How is store secured when closed?

E. Credit Union and Petty Cash

1. Where is physical location of credit union? Plot location on plan.
2. How are records secured?
3. Is a 3-way combination safe available? Is it secured to floor or wall? Does it have rollers or wheels?

4. How much cash is available in the facility? Where is it stored? Plot on plan. How is it secured?
5. When is the credit union open for business?
6. How is the credit union area secured?

F. *Restricted Operations and Areas.*

1. Is government classified material of any type on hand? if so, refer to *Industrial Security Manual for Safeguarding Classified Information.*
2. Are there R&D or QC operations in progress? Plot location on plan.
3. How does your company classify these operations?
4. Are or can the R&D and QC operations be secured from the rest of the immediate area?
5. Is there any area for pilot operations? Plot location on plan.
6. How are these operations secured?
7. Are there any finished product display areas? Plot location on plan.
8. Are the product displays considered restricted until certain dates?
9. Are there areas such as model shops in furniture plants which require a higher degree of security from time to time? Where located? Plot location on plan. What periods of time are involved?

G. *Past Theft Experience*

1. Examine in detail the incidence of theft or pilferage, in the following areas, if applicable:

   a. Finished products or raw material
   b. Company tools (are they marked?)
   c. Resalable or reusable scrap.
   d. Personal property of employees
   e. Employees' automobiles
   f. Vending machines or money changers
   g. Company store

   h. Office machines
   i. Records, plans, blueprints, etc.
   j. Safety equipment.

2. Has any pattern been detected?
3. Are any individuals or groups suspected?
4. Who has investigated thefts in the past?
5. Have any undercover operators ever been used in the past?
6. What departments were they employed in?
7. Were they successful? What department was involved if successful?
8. How many people?

H. Mail and Parcel Post Operations

1. Where is the postage meter machine located?
2. Does it have a check-signing attachment?
3. How are each of the above secured?
4. Who is responsible to secure the area or the machine when not in use?
5. Where are keys kept?
6. Is surreptitious access to the machine possible during operational periods?
7. Does the mailing machine have a company meter ad?
8. Does the facility ship parcel post packages? If so, examine in detail what is shipped, where the item is wrapped, and by whom.
9. Examine the meter record book for obvious inconsistencies. Are there any?
10. Is someone designated to check the record book periodically?
11. Does the mail clerk take the postage meter to the post office for reset of meter?
12. How is check for more postage processed?
13. Check mail available awaiting delivery. Does it contain any personal mail? List names, if applicable, and investigate.

Perimeter Barrier—Chapter 3

Always use the points of the compass when referring to sections of the barrier; i.e., SW corner, E side, NE vehicle gate, etc. This will enable others reading the report to pinpoint locations discussed.

1. Describe type of barrier, height, design, etc.
2. Describe present condition of entire barrier. Pinpoint hazardous areas, undergrowth, holes, washouts, etc.
3. Check storage and material inside and outside the barrier. Is it a sufficient distance away so it cannot be used as a ladder to breach the barrier? Pinpoint each discrepancy.
4. Are poles or trees close enough to be used as an assist to breach the barrier?
5. Can vehicles drive to within a few feet of the barrier?
6. Do any building walls form a part of the perimeter? Study all openings, if applicable.
7. Do storm sewers, utility tunnels, sidewalk elevators, coal chutes or any other openings 96 square inches or larger breach the barrier?
8. List each gate and door opening on the perimeter barrier. Study in detail its use, whether operational, emergency, contractor use, railroad, etc. When studying gates, determine whether single swing, double swing, rolling, electrically operated, or turnstile could better be utilized.
9. Can all gates be adequately secured? Are seals used?
10. Are all gates in good state of repair or are there gaps between gates and fence, gaps under gates, etc.?
11. How is each gate controlled when open?
12. Are chains and locks adequate? This pertains to railroad gates in particular.
13. Are any perimeter barrier anti-intrusion devices installed or CCTV used? Could these devices be used?
14. Do guards patrol the barrier? Should patrols be established?
15. Are there any areas along the barrier where the height of the fence should be increased?

16. Where the fence adjoins a building, is it necessary to increase the height?
17. Where buildings form a part of the barrier are the roofs low enough to be easily scaled?
18. Are keys to padlocks on emergency gates located some distance inside the fence? Are they checked regularly by the guard?
19. Is there a need to enclose attractive nuisances—for instance, ponds and settling basins—with a barrier?

Area Security—Chapter 4

1. Are outside areas free of undergrowth and weeds?
2. Is outside storage organized?
3. Are outside areas free of debris, salvage, scrap, etc.?
4. Where scrap metal is salvaged, can it be adequately secured?
5. Are parking lots physically separated from the manufacturing facility?
6. Are parking lots adequate in size?
7. Are parking lots properly marked for stalls?
8. Is the traffic pattern realistic?
9. Are speed restricting devices used effectively?
10. Are parking lots fenced?
11. Are bumpers used to restrict vehicles from parking next to or damaging the fence?
12. Are gates secured between shifts?
13. Are trucks and/or trailers parked within the perimeter?
14. Are they parked outside the perimeter and exposed?
15. Are any personal vehicles authorized inside the perimeter barrier?
16. Where do visitors park?
17. Where is the mail vehicle parked?
18. Do roads or streets outside the facility present a traffic problem?
19. Are speed limit signs used?
20. Are the interplant roadways adequately marked?
21. Are stop signs and speed limits installed where needed?
22. Are outside area guard patrols established?

23. Are they foot or vehicle patrols?
24. Do they cover the entire area and barrier effectively?
25. Are there any outside installations which should be further protected by installation of another barrier?
26. Is there a trash dump located inside the perimeter?
27. How far from the plant is the trash dump which is used, if applicable?
28. Are any recreational areas within the barrier?
29. Outside the barrier?
30. Do employees use them during lunch breaks?
31. Can nonemployees use them?
32. Is there any construction in progress within the perimeter barrier?
33. How long will it last?
34. Could it be economically fenced off from the remainder of the property?

The Protective Lighting System—Chapter 5

Major changes involving an increase in the protective system should be effected only after a lighting engineer has been consulted. This study must be conducted during daylight *and* during hours of darkness.

1. What type of lighting is used in the outside areas? Incandescent, mercury vapor, etc?
2. How is the lighting system activated?
3. Is there an auxiliary power source available?
4. Is the perimeter barrier adequately illuminated?
5. Is lighting increased at the perimeter gates?
6. Is lighting at employee or pedestrian gates adequate for identification purposes?
7. Are outside storage areas adequately lighted?
8. Are parking lots and truck parks adequately lighted?
9. Are shipping and receiving docks adequately lighted both on the dock and under the trailer?
10. Are any lights objectionable to patrolling guards?
11. Do they expose or blind the guards?
12. Are critical installations such as pumphouses, powerhouses, or suction tanks adequately lighted?

13. Are waterfronts and ship docks adequately lighted?
14. Is lighting increased substantially where little or no barrier exists?
15. Is there any portable emergency lighting available, or is there a need for it?
16. Is there sufficient lighting so guards patrolling have effective observation inside the buildings?
17. Are guards equipped with the proper hand lights?
18. Do security guards report light outages?
19. Are reported outages replaced or repaired the following day?
20. Visualize that lighting you are considering has been installed. Do any heavily shadowed areas now exist which would permit unobserved surreptitious activity?

Building Security—Chapter 6

1. Is a locking schedule established for each building which can be secured? Study it. When, by whom, etc.
2. Check each type of door for security, adequacy of construction, locking device, locking schedule, and controls.
   a. Employee entrances
   b. Emergency exits
   c. Dock doors—overhead and personnel
   d. Doors separating warehouse and production areas
   e. Storage and toolroom doors
   f. Doors to hot working areas
   g. Doors between office areas and manufacturing
   h. Doors between working and idle departments
   i. Doors to danger areas (high voltage, explosives, etc.)
   j. Doors to usually restricted areas (phone equipment rooms, etc.).
   k. Doors to specially restricted areas (R&D., QC, classified operations, etc.)
   l. Doors to "hot item" storage areas (OS&D, etc.)
   m. Doors to walk-in vaults, money rooms, etc.
   n. Boiler room entrances and exits.
3. Study each of the following areas for specific security requirements.

   a. Cafeteria
   b. Company sales store
   c. Credit union office and petty cash handling
   d. Research and development area
   e. Quality control area
   f. Pilot operations area
   g. Display areas
   h. Office and administrative areas.

4. Are employees given lockers?
5. Who furnishes the locks?
6. How are keys controlled?
7. Are locker inspections conducted?
8. Check individual windows for security required on each.
9. What type of and where are anti-intrusion alarms installed?
10. Are they adequate and effective?
11. Can any such devices be used? (See Chapter 11.)
12. Do remotely located emergency exits have any type of deterrent alarms installed?
13. Are they effective? Test for proper operations.
14. Where is the personnel office located?
15. Would it increase security to relocate it elsewhere?
16. Should any activities be relocated for security reasons?

Security of Shipping and Receiving Docks—Chapter 7

Locate all shipping and receiving docks, areas, and warehousing operations on the plot plan.

1. What are the exact hours of operation of the truck docks and railroad docks, and how many days per week are they worked?
2. Are other than shipping and receiving operations performed at the dock? If so, describe.
3. Are guards patrolling the dock areas?
4. Is CCTV used in the dock areas?
5. Are trucks or railroad cars sealed when loaded?
6. When unloaded?
7. What is the sealing procedure?

8. Is the housekeeping at docks adequate to deter pilferage?
9. Are trucks company fleet trucks or commercial carriers, or a combination of both?
10. Is there a waiting room or area for drivers designated, and do they remain there?
11. Do drivers help load their vehicles?
12. Do drivers use the cafeteria, lunchrooms, or rest rooms some distance from the docks?
13. Are the routes to these areas marked?
14. Is there a door or doors designated as a drivers' entrance and is it used as such?
15. Examine the drivers' entrance and physical location of the dock office. Should relocation be considered?
16. Is there any control over the drivers' entrance?
17. Is there a "Will Call" window or desk at the docks? Examine the route used by customers and the method of receiving material.
18. Do customers wander around the docks?
19. Can or should the "Will Call" area be relocated?
20. How are "hot" items secured?
21. Is the area adequately secured?
22. Can or should it be relocated?
23. Are other than dock employees restricted from the dock area?
24. Examine administrative controls and determine if they are adequate?
25. Who checks items being loaded or unloaded to insure accurate count?
26. Is the finished goods warehouse easily accessible from the docks? Can pickers and packers physically be in contact with dock employees and drivers?
27. Is high value freight "buried" on the load, if this is possible?
28. Are incoming trucks with high value freight unloaded immediately?
29. Are damaged freight, broken cartons, etc., recouped immediately?
30. Are they marked with date and quantity number?

31. Check drivers' rest rooms, locker rooms, sleeping quarters, and lounges for evidence of stolen property.
32. Can improvements be made to further deter thefts?
33. Are empty crates, cartons, etc., immediately removed from the dock?
34. Is it feasible to use padlocks on loaded trucks as soon as they are loaded?
35. Are trucks in the park padlocked?
36. Are aisles on the dock kept clear of obstruction?
37. Is the security seal procedure, as established by policy, being followed?
38. Are seals adequately secured and accounted for?
39. Are seals checked for tampering or other discrepancies on arriving trucks?
40. Are all units on the lot sealed, locked, ignition keys removed, etc., per the procedure established?
41. Are freight bays clearly marked and identified?
42. Is freight stacked properly?
43. How is trash accumulated at the dock removed?
44. Are trash barrels located on docks?
45. Does the trash removal procedure make it possible to commit thefts in its removal?

Locks, Key Control, and Security Containers—Chapter 8

Examine all keys, records, and key containers in existence.

1. Who is responsible for key control?
2. What type key box is used and how is it secured? How are records kept?
3. Is record keeping adequate?
4. What should be done to regain control if it has been lost?
5. Is there a record of individual locking device installation?
6. Examine each locking device in use. Are they adequate to secure that particular area?
7. Are they in good working condition?
8. By whom are records inspected?
9. When are key inspections conducted?

10. What type locks are generally in use? (Removable core, many different makes?)
11. How are duplicate keys secured?
12. If keys are cut at the facility examine security of cutting machine, key blanks, and supply of pins and springs. Is security adequate?
13. Are keys identified by name or codes stamped on the bows?
14. Are locks used to secure special restricted areas changed periodically? Is it necessary?
15. Does the system consist of a grand master, master, and submaster keys?
16. Examine issuance of grand master and master keys. Are these issued only on a need-to-have basis?
17. What is done if any class of key is lost or compromised?
18. Are all combination locks the three-position type?
19. Are movable safes located where patrolling guards can observe them?
20. Are they illuminated at night? Are they chained or otherwise secured against easy removal?
21. Determine who has knowledge of combination of each container with this type of device. Do these persons have an actual need to know?
22. Are the combinations recorded and stored elsewhere?
23. Is the security of these combinations adequate?
24. Test persons for combinations. Ask what the combination is to a particular container.
25. Ask the person where he or she keeps record of combination.
26. Who changes combinations and how often?
27. Are combinations changed when personnel who have knowledge of them leave?
28. Do guards issue any keys on a temporary basis?
29. How are these controlled?
30. Is issue restricted to certain individuals?
31. Examine each security container used to store the following critical items and determine if each is adequate:

    a. Cash, whatever the amount. Check signing machine
    b. Records—administrative, payrolls, etc.

c. Records and plans—company classified R&D, etc.
d. High-value metals—gold, silver, nickel, etc.
e. Government-classified material—see DOD manual.

32. Review your study and analysis of locking devices on individual doors. Are any further changes necessary?

Identification and Control of Personnel and Vehicles—
Chapters 9 and 10

Study the application form for sufficiency of information. Study the investigation report forms for completeness.

1. Should changes be made?
2. To what extent are background investigations of all categories of personnel conducted?
3. Are ID badges or cards presently in use?
4. Is there a need for color-coding?
5. Are ID cards issued to employees on a permanent basis?
6. Are they worn in a uniform manner?
7. Are they displayed on outer clothing when reporting to, while at, and when departing the facility?
8. What procedure is followed when an employee loses the identification media?
9. When it is forgotten?
10. Is paper stock used to produce cards adequately secured?
11. Does the identification contain at least:

    a. Color photograph
    b. Name and social security number, title, position, or department
    c. Card serial number
    d. Name of facility
    e. Signature of bearer
    f. Signature of person authorized to authenticate and issue cards?

12. How are visitors identified and controlled? Is the control adequate?
13. How are outside contractors controlled? Is the procedure adequate?

14. How are outside vendors controlled? Is this adequate?
15. How are applicants for employment controlled? Is this adequate?
16. How are longshoremen and ships' crews controlled?
17. Are routes through the facility marked?
18. Study the location and number of employee entrances, number of employees using each entrance, and what controls are in force. Are present controls adequate?
19. Where are time clocks located?
20. Is it possible, for better control, to group all clocks together?
21. Is it possible for employees to punch others' cards?
22. What other special entrances for other than employees exist? Who uses them, and what controls are in effect?
23. Are controls of employee flow to and from the company store adequate?
24. Are fire wells or emergency exits used for operational purposes?
25. Does this create a security hazard?
26. If the facility uses elevators—passenger or freight—how are they controlled against unauthorized use?
27. Are they automatic or manually operated?
28. If manual, who operates them?
29. Examine the routes used by employees from the entrance to the locker rooms and then to the various work areas. Are material, goods, etc., exposed to theft?
30. What changes should be made?
31. Should entrances be changed or others created?
32. Are any personnel presently using unauthorized entrances or exits?
33. What changes should be made?
34. Are groups authorized to visit the facility?
35. How are they controlled?
36. Are all personnel in a group accounted for at the exit?
37. Do routes used for groups unnecessarily expose company property to theft?
38. Are employees issued uniforms? If so, how are they issued, cleaned, and stored?

39. Is color-coding of uniforms practical to increase control?
40. Study controls exercised over employees during coffee and lunch breaks in each department. Can employees eat at their work stations?
41. Do all employees leave their department, the warehouse, or docks during lunch?
42. Can some stay? Examine each area closely—this is a critical time, when thefts can easily be committed.
43. How are employees who leave during shifts because of sickness, emergencies, etc., controlled?
44. Are controls relaxed over those enforced at shift changes?
45. Could employees feign sickness to take advantage of relaxed controls and perpetrate thefts?
46. Are any of the following registers used at points of ingress and egress? If so, examine each to determine if sufficient information is being provided; they are being properly completed; they are inspected or examined regularly; they are being filed at least temporarily to facilitate investigation if the need exists.

    a. Visitors' register
    b. Employee-lost or forgotten ID card register
    c. Vehicle registers—vendor and company fleet.

47. Are buildings and interplant roadways marked well enough to insure that authorized persons not familiar with the facility do not become lost?
48. If not, how could this be improved?
49. Are employees authorized to park within the protected area? If so, is or can the area be effectively fenced off?
50. Are any other private autos authorized to park within the perimeter? Whose cars, where, during what hours? Is it possible to move them elsewhere, if they create a hazard to security?
51. Are vehicle identification decals being used?
52. Are they adequate?
53. Who issues them?
54. If they are not used, should they be?

55. Would color-coding of decals assist in more efficient parking and control?
56. How do guards become involved in employee parking? (Do they direct traffic or police the lots? Do they open and close gates?)
57. Is the traffic pattern in the parking lot established to direct a smooth, safe flow of vehicles? Are any changes obviously required?
58. Where do vending machine servicemen, cafeteria supply trucks, mail trucks, utility repair trucks, etc., park?
59. Could any of the established procedures be a source of theft?
60. Are truck cabs and closed bed trucks inspected by the guards?
61. In addition to being registered in and out, is any type of vehicle identification used?
62. How are outside contractor vehicles controlled, and do they use a special gate?
63. Are vendor servicemen's vehicles inspected upon departure?
64. How are railroad personnel controlled? Do they have free access inside the facility?
65. Are departing railcars regularly inspected?

Fire Protection and Prevention—Chapter 13

1. Does the facility have a sprinkler system installed in all areas?
2. Study the system. Is it a wet and dry combination?
3. How many risers comprise the system?
4. How many fire zones?
5. Are risers monitored with electrically operated water flow alarms?
6. Where are they monitored?
7. Are risers numbered?
8. Are riser locations marked so they are not inadvertently blocked?
9. Are any blocked?
10. Could riser locations be protected against damage if guard rails were installed?

11. Who inspects and tests the system?
12. How often and how are tests conducted?
13. Are all OS&Y valves open and sealed or alarmed?
14. Do dry system risers have heated riser houses?
15. Are they clean and free of debris?
16. Are there spare sprinkler heads, wrenches, and plugs at each riser location?
17. Do hand extinguishers have inspection tags attached?
18. When were they last inspected?
19. Are extinguisher locations adequately marked?
20. Are extinguishers clean, not used as clothing hangers, and not blocked?
21. Are there any inoperable or damaged extinguishers?
22. Are extinguishers plainly marked by class?
23. Are they given a numerical designation and the location given the same number to insure the proper extinguisher is returned?
24. Can the extinguisher location be seen from all points in the area it is intended to serve?
25. Are extinguishers of the proper type for the area?
26. Are spare extinguishers available as replacements?
27. Is there an area designated where used extinguishers are placed?
28. Who services the extinguishers?
29. Are all extinguishers properly hung or stored?
30. Question some employees regarding location of nearest extinguisher, its type, and how it is operated. Are they knowledgeable?
31. Are hand-drawn extinguishers parked in protected areas?
32. Are they properly marked?
33. Are hoses on these extinguishers in good condition?
34. Examine all fire doors. Are they operable?
35. Are fusible links in place?
36. Are weights and counterweights enclosed?
37. Are guards installed to prevent doors from being blocked?
38. Pinpoint all blocked or inoperable doors and list repairs needed.

39. Are there any doors which are used to secure flammable storage and which are not covered by or constructed of metal?
40. What type of trash cans are used?
41. Are they constructed of metal?
42. Do they have self-closing tops?
43. Are flammables stored in unprotected areas?
44. Examine housekeeping of all areas of the plant with particular attention to locker rooms, boiler rooms, electrical equipment rooms, flammable material storage areas, etc. Inspect basements, lofts, and firewells. Note conditions.
45. Is smoking prohibited in certain areas?
46. If so, are the rules being broken?
47. Are there manual-pull boxes located throughout the facility?
48. Do building occupants know their location?
49. Is there an audio alarm system installed throughout the facility?
50. Is it coded by area?
51. Should it be coded?
52. Is there a standpipe hose system installed?
53. Are all hoses properly hung or stored?
54. Are they properly located?
55. What is the condition of the hose upon visual inspection?
56. Are yard hydrants free of weeds and debris?
57. Are hose houses located over hydrants or could they be?
58. Are hose houses sealed or locked with breakaway locks?
59. What is the condition of each hose house? Is necessary equipment available?
60. What is the condition of the hose and other equipment?
61. Are post indicator valves open, sealed, and free of weeds and debris?
62. Are hydrants or post indicator valves located near roadways or operational areas adequately protected?
63. Are there gas pumps at the facility?
64. If required, are they protected?
65. Is no smoking in the area enforced?

66. Is spillage immediately washed or otherwise removed from the area?
67. Where are propane or gasoline fueled industrial vehicles fueled?
68. Are safe practices in effect?
69. Is the location outside the principal buildings?

Safety for Personnel—Chapter 14

As you conduct your survey, note whether required safety equipment is being worn and make notes of specific areas, instances, etc., where violations of rules were observed.

1. Is a safety director appointed?
2. Is there a safety program and what does it consist of?
3. How often does the safety committee meet?
4. Who is on the committee?
5. Are safety shoes required?
6. Are safety hats required?
7. Are safety gloves required?
8. Are safety glasses required?
9. Are safety aprons required?
10. Are full-time nurses or doctors on duty?
11. Should they be?
12. How are injured employees treated by in-plant first-aid people and who are the first-aid people?
13. How many are available on each working shift?
14. Are guards trained to administer first aid, and when do they?
15. Are they qualified?
16. Are ambulances or emergency vehicles available at the facility?
17. Who is responsible for their operation on each shift?
18. Are operating personnel properly trained?
19. Are safety posters used?
20. Where is the first-aid room located?
21. Can it be adequately secured?
22. Are the cabinets in the room adequately secured?

23. Is there a place for the injured to lie down?
24. Should it be moved to a more central location?
25. Are there any narcotics on hand?
26. Are they properly secured and inventoried?
27. Are they dispensed only on doctor's prescription?
28. Are prescription files maintained?
29. Are visitors authorized in restricted areas required to wear safety equipment required in that area?
30. How is this equipment, when issued, accounted for?
31. Examine condition of all floors, stairwells, handrails, stairs, ladders, etc., for obvious safety hazards.
32. Does all machinery have guards where necessary?
33. Are the guards and safety equipment installed on machinery being used as intended?
34. Are mirrors used on blind turns?
35. Would yellow flashing lights increase traffic safety of industrial vehicles inside the plant?
36. Are forklifts and other industrial vehicles operated recklessly?
37. Are the traffic aisles properly marked?
38. Are chocks used on parked trailers?
39. Are speed limit signs posted along outside industrial thoroughfares?
40. Could speed reducers be effectivey used? Where?
41. Do employees use excessive speed in parking lots?
42. If so, how should speed be controlled?
43. Does traffic to and from the plant to the public street or highway present a safety factor?
44. How can the hazard be reduced?

Further Aspects of Theft and Pilferage Control—Chapter 15

1. Are lunch-box inspections being conducted?
2. Is a package pass system used?
3. Are lockers available at point of entrance, for store employees' packages?
4. Are specimen signatures available for guards?
5. Is there a box available near the exit where employees could deposit company items they carry out "by mistake"?

6. Are tools loaned to employees?
7. If so, how are they controlled?
8. Study removal of scrap or salvage in detail to determine if controls are adequate.
9. Study removal of all trash in detail to determine if controls are adequate.
10. Do guards inspect outside areas and the perimeter barrier for property which has been hidden?
11. Do guards ever supervise or observe trash removal? Do they accompany the trash trucks?
12. If the facility has gas pumps, examine control procedures closely. Are they adequate?
13. Can pumps be locked and are they?

Emergency Evacuation and Disaster Planning—Chapter 16

1. Are emergency evacuation and disaster plans formulated and reducted to writing?
2. Are they up-to-date?
3. Are emergency exit doors adequately marked?
4. Are exit lights operating?
5. Can emergency exit doors be opened without use of a key?
6. Are they blocked?
7. Are emergency evacuation plan drills conducted?
8. Do employees know the routes to and the exit doors their department should use?
9. Are emergency evacuation aisles plainly marked?
10. Are they kept free of obstacles?
11. Are outside assembly areas designated and marked?
12. Are they a sufficient distance from the buildings to insure that they do not expose employees to possible injury or block emergency equipment routes?
13. Are guards included in the emergency evacuation and/or disaster plan procedures?
14. Are the assigned responsibilities commensurate with the over-all abilities of the guard force?
15. Do guards have an up-to-date list of emergency phone numbers?

16. Would their calling interfere with other designated duties?
17. Could the triangular alert system be more effectively used?
18. Are civil disturbance or bomb threat plans formulated and in writing?
19. Are they necessary in this particular area?

Index